a FUTURE HISTORY *of* WATER

a Future History

Duke University Press Durham and London 2019

of Water

Andrea Ballestero

Printed in the United States of America on acid-free paper ∞
Designed by Mindy Basinger Hill
Typeset in Chaparral Pro by Copperline Books

Library of Congress Cataloging-in-Publication Data
Names: Ballestero, Andrea, [date] author.
Title: A future history of water / Andrea Ballestero.
Description: Durham : Duke University Press, 2019. | Includes bibliographical references and index.
Identifiers: LCCN 2018047202 (print) | LCCN 2019005120 (ebook)
ISBN 9781478004516 (ebook)
ISBN 9781478003595 (hardcover : alk. paper)
ISBN 9781478003892 (pbk. : alk. paper)
Subjects: LCSH: Water rights—Latin America. | Water rights—Costa Rica. | Water rights—Brazil. | Right to water—Latin America. | Right to water—Costa Rica. | Right to water—Brazil. | Water-supply—Political aspects—Latin America. | Water-supply—Political aspects—Costa Rica. | Water-supply—Political aspects—Brazil.
Classification: LCC HD1696.5.L29 (ebook) |
LCC HD1696.5.L29 B35 2019 (print) | DDC 333.33/9—dc23
LC record available at https://lccn.loc.gov/2018047202

Cover art: Nikolaus Koliusis, *360°/1 sec, 360°/1 sec, 47 wratten B*, 1983. Photographer: Andreas Freytag.
Courtesy of the Daimler Art Collection, Stuttgart.

This title is freely available in an open access edition thanks to generous support from the Fondren Library at Rice University.

PARA LIOLY, LINO, RÓMULO, Y TÍA MACHA

CONTENTS

PREFACE

I was walking toward the exhibit in the 2009 World Water Forum held in Istanbul, Turkey, when I heard someone call my name. Surprised, I turned around to see Lucas, a friend from Ceará, in northeastern Brazil, who at the time worked at the Water Management Company created in the 1990s when the state revamped its water institutions. I was happy and surprised to see him. After we greeted each other he told me he had collected a couple of things that I would find interesting and handed me a poster and a brochure he had picked from an NGO in the exhibit I was trying to get to. As I unrolled the poster I was astonished. Without knowing, out of the dozens of stands, Lucas had picked up and was handing me a poster produced by an organization in Costa Rica that I had been following for several years. I thanked him profusely, and after we said goodbye I found myself pondering how all the particularities of location that I had imagined would ground my research had just been troubled. Geography and location were too performative, too flexible to use as grounding devices for my research.

Lucas and his colleagues from Brazil, the NGO, and state representatives from Costa Rica, and I were all fellow travelers in this international water circuit. They all were giving talks about their experiences in shaping the political materiality of water, telling stories about how they were mobilizing categories, challenging legal infrastructures, questioning economic models. Their talks described particular experiments, new attempts to change the future of water, and the specific tools they were using to do so. All their stories were about the possibility of different futures, narrations where the materiality of the present—rivers, water pipes, rain patterns, evapotranspiration rates, land titles, and water pumps—was experienced as an anticipatory event, as a trace of the yet to come.

Without being able to rely on geography to stabilize my research, I quickly refocused on those futures and the technical crafts involved in bringing them about. Understanding the ethical possibilities for the future that they inscribed in their technical craft required me to pay attention to practices and artifacts that often seem unremarkable, or even worse, uninteresting tools of familiar economic and legal systems we wish to undo. In this book I suggest that those knowledge forms and the practices by which they are brought to matter are devices with wondrous capacities to transgress ontological boundaries, even while seeming to merely replicate what currently is. Rediscovering these devices and their wonder reminds us of the intensity by which everyday life, including technocratic life, constantly shapes the limits of the possible.

In philosophical terms, wonder takes over when knowledge and understanding cannot master what they should. It arises when, "surrounded by utterly ordinary concepts and things, the philosopher suddenly finds himself [*sic*] surrounded on all sides by aporia" (Rubenstein 2006). Wonder (*thaumazein*) is regarded as the point of origin of Western philosophy (Socrates/Aristotle). Yet, as with many origins, this one is also imagined as in need of being superseded because of its pathos (Aristotle), its heretical implications (St. Augustine), and/or its lower value as a passion that is closer to the feminine and the childish (Descartes!). For ethnographic analysis, however, the task when thinking with wonder is different. If wonder strikes when people, things, and other beings encounter each other in concrete times and places, the analytic task is to trace how those encounters redefine wonder as an affective disposition. This is what the process of doing the research for this book and writing it did to my own thinking. The four devices I present in this book reshaped the sense of ethnographic wonder with which I embarked on the project. In dry technocratic procedure, I found space for wondrous wonderings. Thus, rather than defining wonder as a particular vision of the world, I want to invite you to think of wonder as an underlying epistemic mood.

In its Western philosophical trajectory, wonder has ended up resembling the concept of marvel or enchantment. But that is not the only meaning wonder has. Wonder is instability, confusion, maybe even frustration. It entails a fluidity that, while rendered enjoyable and desirable in much anthropology, also entails a type of difficulty and disorientation that is not necessarily a pleasurable sensation.

When denuded of its positive valence, wonder is much more textured and less idealized. It entails openness and the potential expansion of possibilities. It is more than the comfortable position of the modest witness, or the point of view from nowhere, or the God trick. It is dirty, messy. It can make you allergic, want to avoid it. From this point of view, one could not limit an anthropological wonder to worlds that differ radically from the liberal tradition (Scott 2013).[1] Social analysis that begins with wonder is moved by a "peculiar cognitive passion that register(s) the breach of boundaries" (Daston and Park 1998: 363), regardless of where those boundaries were originally placed. Wonder opens up familiar worlds for rediscovery. The predecessor of this type of wonder is the early modern collection of oddities and its attempt to reorganize worlds and beliefs.

Surfacing throughout Europe in the sixteenth century, after Christopher Columbus's imperial travels to the Americas, collections of "all curiosities naturall or artificial" began to proliferate in Europe (Hodgen 1964: 114). First put together by the aristocrat, merchant, or eccentric personality, the collection of oddities gathered "books, manuscripts, card-games, coins, giants' bones, fossils, . . . zoological and botanical specimens" (Hodgen 1964: 115). The items in the collection were extraordinary as well as unremarkable. Smaller items were stored in cabinets or cupboards following the design of the apothecary shop. Larger artifacts were suspended from walls or ceilings, enveloping the body of the observer. The result was high density and the accumulation of semiotic charge until it could barely be contained. Due to this aesthetic uniqueness, these collections came to be known as *les cabinetes de curiosités* (cabinets of curiosities) or *Wunderkammer* (cabinets of wonder).

Part of the power of the cabinet of wonder resided in how it took the familiar form of the geologic and botanical collection, repurposed it, and transformed it into something very different. While those collections recorded "natural" taxonomic ontologies, the collection of oddities reconsidered inherited hierarchical structures and the limits of nature. It was a "force-filled microcosm" unlike any other, since each collection was a unique and unrepeatable assemblage (Frazer 1935: 1). Due to this transgressive nature, the collection reinforced a sense of chaos at a time of major cosmological transition, an era when European colonists confronted a world that no longer was what they thought it used to be.[2] By grouping artifacts of radically different origins and forms, collectors challenged inherited orders and made new ones possible. This openness showed the power of setting things side by side in one formation, even if the things brought together did not seem to belong next to each other—a manufactured tool, a doll, and a leaf could all be part of a single heterodox set.

This collecting impulse, and its accompanying sense of wonder, was not limited to artifacts that could be placed inside a drawer. Another type of object, one that did not lend itself to easy placement, was also pursued: the manner or custom. Impossible to hang from a wall or put in a drawer, the custom was suspended on the page of the printed book. It required description, translation, and illustration, and had to be connected to ideas such as nation, society, and civilization. In Europe, the most popular and well known among the early collections of customs was *The Fardle of Façions* by Johan Boemus, translated into English in 1555 (Hodgen 1964).[3] The book describes cultural groups by way of their laws and institutions, including marriage systems, religion, funeral practices, weapons, diet, and apparel (Hodgen 1964: 287). Boemus wrote the book with two objectives. First, he wanted to make accessible to a broader audience existing knowledge about the variability of human behavior. Second, the book was written to improve the "political morality" of his readers and expose them to "the laws and governments of other nations," with the purpose of developing intelligent "judgments" as to the best "orders and institutions" to be fitted into new colonial lands (Hodgen 1964: 131). In today's terms, the book was a collection of case studies, an early modern repertoire of techniques for colonial control so successful that it was reissued at least twenty times and translated into five languages.[4]

Fast forward five centuries, and the collection of customs, with its analogical structure and the wonder it inspires, still prevails as a means to imagine sociomaterial improvement and cultural difference in many circuits, including the World Water Forum. Described as compilations of best practices and policy tools, and brought together in documents such as manuals, frameworks, and anthologies, these contemporary collections circulate nationally and internationally with the purpose of "improving" the "political morality" of water. These documents juxtapose "models" from different countries, environments, and societies to offer possible answers to collective questions, such as how to improve community participation in water management, how to charge just prices for water services, or how to guarantee the human right to water for all.

And also just like Boemus's, these collections are not cohesive arguments about the proper, but heterogenous samples of the possible. Their constituting items can contradict, complement, expand, or oppose each other, and yet the collection remains viable as a summation of items that preserves their odd asymmetries. This book replicates that epistemic gesture. It takes you into a particular collection of devices, into their histories and the actions by which they are activated to produce what the professionals among whom I worked see as the necessary ethical bifurcations to transform a world that always resists change.

The curatorial work behind this collection takes "odd" technocratic devices that we often take for granted and suspends them on the page of the book. The devices I bring together come from different parts of the world and are not homologous in any way. Each is a microcosm of selected histories and possible futures that conveys an expansiveness that is difficult to capture. At the same time, each device gives the sense of being a thing in and of itself. But just as with the premodern collection of oddities, what I want to emphasize is how, when we put them together into a collection, these devices invite us to wonder about what we take as self-evident. I imagine this book as an invitation to linger in wonder, as we encounter familiar worlds.

ACKNOWLEDGMENTS

Books are collective accomplishments. People, organizations and companion species make them possible. Those listed below had a direct impact on the book itself. Many more helped bring about the project in the first place. I am truly thankful to all of you in so many different ways. I hope you can see some of yourselves in the pages that follow, I certainly can.

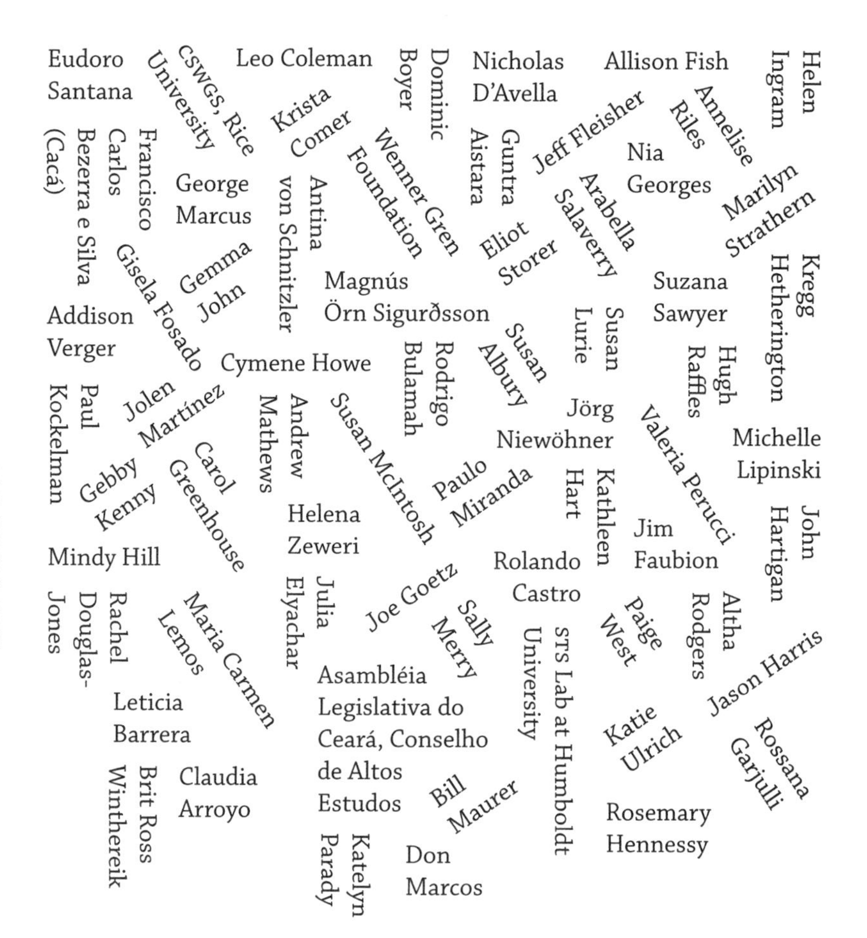
Eudoro Santana
CSWGS, Rice University
Leo Coleman
Dominic Boyer
Nicholas D'Avella
Allison Fish
Helen Ingram
Krista Comer
Jeff Fleisher
Annelise Riles
Francisco Carlos Bezerra e Silva (Cacá)
George Marcus
Antina von Schnitzler
Wenner Gren Foundation
Guntra Aistara
Nia Georges
Arabella Salaverry
Marilyn Strathern
Eliot Storer
Gemma John
Gisela Fosado
Magnús Örn Sigurðsson
Suzana Sawyer
Kregg Hetherington
Addison Verger
Susan Lurie
Susan Albury
Hugh Raffles
Cymene Howe
Rodrigo Bulamah
Jolen Martínez
Paul Kockelman
Andrew Mathews
Susan McIntosh
Jörg Niewöhner
Valeria Perucci
Michelle Lipinski
Gebby Kenny
Carol Greenhouse
Paulo Miranda
Kathleen Hart
Helena Zeweri
Jim Faubion
John Hartigan
Mindy Hill
Rolando Castro
Rachel Douglas-Jones
Maria Carmen Lemos
Julia Elyachar
Joe Goetz
Sally Merry
Paige West
Altha Rodgers
Jason Harris
STS Lab at Humboldt University
Leticia Barrera
Asambléia Legislativa do Ceará, Conselho de Altos Estudos
Katie Ulrich
Rossana Garjulli
Claudia Arroyo
Brit Ross Winthereik
Bill Maurer
Rosemary Hennessy
Katelyn Parady
Don Marcos

INTRODUCTION

Around noon on the fourth day of the World Water Forum, held in 2006 at Mexico City's convention center, fifty out of the ten thousand participants managed to sneak in the necessary tools to stage a surprise protest. As the demonstrators went through the metal detectors that turned entry doors into security checkpoints, the guards inspecting their personal belongings ignored the water bottles, small coins, and folded pieces of cloth that they were bringing into the building. The would-be protestors walked briskly toward the lobby, where three levels of meeting rooms connected through an intricate system of balconies and escalators, creating an ideal stage for attracting an audience. Within minutes, empty plastic water bottles emerged, coins were dropped into them, and cloth signs unfurled. The protestors began shaking their bottles rhythmically and chanting: *El agua es un derecho, no es una mercancía! El agua es un derecho, no es una mercancía!* (Water is a right, not a commodity! Water is a right, not a commodity!)

With the opposition between a right and a commodity, the demonstrators were not invoking just any right; they were referring to the human right to water. Their voices were tactically recruiting water's universalism to denounce the injustices and dispossession occurring around the world as a result of its commodification. Their chant was more than a mere demonstration slogan; it was a calculated rhetorical move marking the practical and material distinctions between human rights and commodities. The demonstrators were convinced, as were many other participants in the forum, that water should be a universal human right accessible to all, and for that reason should never be commodified. But they also knew that those distinctions need to be produced in all sorts of places; courts were not the only spaces where rights were enacted, and markets did not hold a monopoly over commoditization practices.

Figure I.1. Disposable water bottle turned protest rattle.

The sound of the shaking bottles in the protestors' hands immediately attracted security guards, who approached from all corners of the building and threatened to detain them unless they stopped. After heated exchanges, the protesting voices slowly quieted and the plastic-metallic rattling of the bottles stilled. What had been a hub of intense energy dissolved, quickly reverting to the hum of a controlled, professional environment. If you had entered the lobby at that moment, you would not have imagined a vigorous protest had just ended. The significance of that historical moment had become precarious—a happening whose energetic exuberance had been effaced.

Among all of the things one might find intriguing about this protest, the shaking bottles are what continue to captivate me so many years later (see figure I.1). Inhabiting the space previously occupied by water, the coins inside the bottles insinuated that water had been transubstantiated into money, the ultimate commodity. While the demonstrators' chant created a clear structural bifurcation between human rights and commodities, the coin-filled bottles confounded the clarity of that contrast. With their rhythmic movements up and down and the penetrating sound of metal pounding against plastic, they complicated the clarity of the protestors' words. These bottles were sound-making instruments and statements about water's confounding nature. They were conceptual things, material abstractions.

These protest bottles, with their unruly embroilments, became the conceptual locus of my research on the technolegal politics of water. What kind of relation was there between the activists' words, with their clear partitions, and the bottles in their hands, with the transubstantiation they suggested? If water is to be a human right, and not a commodity, how do you differentiate these two legal and economic formulations? And more generally, how do people create distinctions and bifurcations if the world in which they live constantly drifts toward entanglement, blurring stark oppositions?

These questions are not only relevant to our thinking about the politics of water, they go beyond. Human rights and commodities directly shape or distantly hover over much of the organization of value, collective life, and nature. The relations between property and body parts, health and healing, food, nature, and even access to the internet are all discussed through similar oppositions: should they be human rights or "just" commodities? As we see, commodities and human rights are generative ethnographic objects; they are classifications already shaping the world. From theological discussions of natural rights, to moral arguments about property, all the way to the universalisms that defined human dignity in the twentieth century, these two notions continue to establish the conditions of possibility for life and death in the twenty-first century. Not surprisingly, however—as the shaking bottles teach us—what from a distance seem to be clearly distinct ideas, under closer inspection are far from that. For example, does paying for water automatically turn it into a commodity? Is the collective responsibility to care for water enough to transform it into a human right? Can a legal definition transform a commodity into a human right?

This book is designed to address the nuances of these questions. I conducted most of the fieldwork for this project in two Latin American countries: Costa Rica and Brazil. I selected these sites because these two countries were among the few in the region that had not formally incorporated an explicit recognition of the human right to water into their national laws or constitutions. This omission created a climate of ongoing struggle among the activists, experts, and public officials I worked with. Their struggles included the promotion of legal reforms, creating more just water pricing systems, and experimenting with more democratic water management programs. Given that they could not fall back on the symbolic power of the law to promote the human right to water, they took those processes as opportunities to affirm the distinctions they are committed to, the distinctions between a human right and a commodity. This book centers on that work and examines the affective, epistemic, and political work of making distinctions matter.

The people with whom I worked in Costa Rica and Brazil devote their energy and time, and sometimes even their lives, to creating a difference that matters, a separation that they hope will make clear what practically, and even morally, sometimes seems blurred.[1] They do that work from a variety of locations: NGOs, bureaucratic offices, scientific institutions, and even their respective congresses. They are economists, lawyers, engineers, environmental scientists, philosophers, sociologists, farmers, schoolteachers. They consider their technical work—a combination of legal, economic, and hydrologic knowledge—a tool to attain ethical goals. For them it is not sufficient to state that water is a human right, as if the mere act of placing it under a general category accomplishes the outcomes they hope to achieve. They are interested in what exactly that difference means and for whom, what forms of collective life are implicated by creating a distinction. But this does not mean they are all in agreement. My interlocutors hold different political ideologies, represent contradictory interests, and have built their political and technical authority on their active involvement in or opposition to policy-making efforts. At the same time, they are all active participants in national and international networks, such as the World Water Forum, where people share the latest frameworks for action and participate in training workshops and technical talks.

Since 2003, I have talked to this group of activists and experts in their offices, on field trips, at workshops and community meetings, and in many other settings where they have had to articulate for themselves and others how they define the difference they want to see in the world. I also met with them in other countries where we were all attending international water meetings, such as the World Water Forum. I conducted interviews and fieldwork in Spanish, Portuguese, and English. To prepare for our conversations, I had to learn about the technical dimensions of their ideas, which in turn required delving into legal doctrine, economic theory, and organizational techniques. Across those different locations and areas of knowledge, my interlocutors always brought me back to the question of how a human right and a commodity are made different. They emphasized that to act in the world is to change the future by defining differences that are ethically important.

This book centers on the imaginative work they do to create these valued distinctions. I analyze the work necessary to separate categories that resist separation—a condition that is experienced by all sorts of people around the world, anthropologists included. Following what the protestors and their shaking bottles taught me, my analysis does not take us to the usual locations. I do not trace human rights in courts or commodities in markets. Instead, I follow water activists and experts as they attempt to create those separations across other kinds of locations: cubicles, community meetings, international workshops, and even Excel files. Throughout those locations, they attempt to produce the preconditions of futures where differences become plausible and entanglements do not preclude the viability of the distinctions necessary for a more just form of sociality. Through that work we will see how water is kept mattering through the everyday bureaucratic and technical decisions whereby its very materiality is at stake. Through that work we can also understand how people connect their everyday work to a future that has not yet arrived. The chapters in this book focus on the assumptions imbued into the technical tools through which the work of differentiation is performed; they show how people touch the future with their technolegal tools. I specifically focus on four instruments people use: a formula, an index, a list, and a pact. I show how each participates in making the future history of water while attending to how these technolegal tools have become staples in the organization of all sorts of legality and authority (Johns 2016). As I show, these tools quietly determine the limits of the possible by both narrowing down certain options and opening the possibility of creating different, and maybe better, worlds. This book attends to that dual potential and this introduction elaborates on the conceptual work that potential requires.

BIFURCATIONS

As I conducted fieldwork for this project, I became more and more captivated by my interlocutors' commitment to create distinctions despite the slipperiness of the worlds they were part of and the slippages between the concepts that guided their work. Thus, I came to see the differentiations they worked for as forms of bifurcation, "moments when terms cannot be taken as self-evident and require explicit reference [not only] to their meaning" but also to their semiotic tensions with other terms (Strathern 2011).

I find two things particularly helpful in the idea of a bifurcation. On the one hand, it shows how things that seem to be unitary are in fact separations waiting to happen (see figure I.2). On the other hand, the notion of a bifurcation reveals that once a first separation has been produced, if we continue looking, we realize that what seems to be just one of two is in fact an already entwined line requiring a new differentiation, a new bifurcation. In the world of water, for example, it looks like this: if regulators decide they will keep the price of water tied to inflation to make it a human right, once they have performed that operation they still have the problem that water continues to be a commodity people are paying for. Thus, they need to perform a new differentiation to affirm, in some other way, its humanitarian nature. Following the lines in figure I.2 makes this dynamic visual.[2] This never-ending bifurcating mesh reveals that there is no end point to this kind of work: once a bifurcation is effected, a new one becomes necessary for each of its branches. Thinking about making differentiations in the world in this way emphasizes that such processes occur in time, as ongoing attempts that are never fully finalized.

Keeping things clearly separated and distinct has important consequences (see also Candea et al. 2015; Roberts 2017). In the cases I studied, making things distinguishable helps people decide whether a water valve is legally closed, what kind of price increase would preclude profiting from water, and who is held responsible for water supply at times of scarcity. But as soon as those separations are successfully put in place, what was clear blurs, revealing unexpected consequences that seem to undo the clarity people like my interlocutors worked hard to achieve. It is as if the separations they put in place are political and moral arguments that "take off in one direction by rendering another [direction] also present" (Strathern 2011: 91). Because of this dynamic, the bifurcations they produce are a mesh of distinctions that sidestep any simplistic dualisms; the only clear-cut effect a bifurcation produces is the need to determine new and future distinctions.

The time I spent with my interlocutors showed me firsthand how the world of bifurcations operates. Converting water into a human right entailed keeping the implications of its commodification at the forefront; arguing for its commodified exchange depended on mobilizing humanitarian logics of universal access. In this kind of bifurcating mesh, a human right and a commodity are absent presences to each other, figures that shape each other's respective forms from within and preclude any easy reduction-

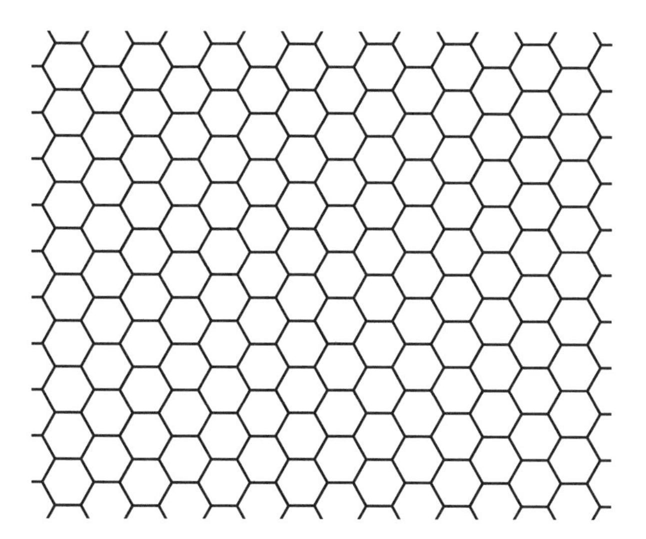

Figure I.2. Mesh of never-ending bifurcations.

ism. It is in this situation that my collaborators' work becomes a constant effort to make distinctions recognizable, since the more you try to clarify and separate, the more you bring about mutuality. As I will show, these differentiation struggles turn water into a planetary archive of meaning and matter (Neimanis 2012: 87), an archive that is constituted through ongoing processes of abstraction and materialization where word and matter, formalization and substance, are inseparable (Barad 2003; Helmreich 2015). But there is more. As I will show, it is through these processes that people like my interlocutors are quietly and constantly elucidating profound questions about the meaning of life, property, and subjectivity at the beginning of the twenty-first century—a time when science has diagnosed the Earth as being already anthropogenically transformed and when the notion of the Anthropocene occupies those with a planetary imagination.

Making differences is not an easy or innocent task, though. The water professionals I worked with create these differences from a subject position that is far from any idealized modern imagination of the individual as the master of history. The dream of *Homo faber*, as the fabricator of the world bringing permanence, stability, and durability (Arendt 1959: 110) to make events match her desires has been long dissolved, if it was ever there at all. Inherited and long-standing economic asymmetries, the inertia of legal systems too baroque for their own good, a bureaucracy that moves extremely slowly, and all-too-uncontrollable environmental events quickly dissolve any sense of control to produce radical transformation; there is too much path dependency and too much recalcitrance (see also Riles 2013). In place of that maker of linear histories of cause and effect, we find a humbler figure whose capacity to act is directed toward tactical modifications—transformative shifts that are unpredictable. This subject locates the possibility of change not in a historical metanarrative but in the concrete junctures where she conducts everyday political and epistemic labor to effect bifurcations. These junctures include things like a legal definition, a percentage, a variable in a formula, or a promise. I conceptualize each of these junctures as a technolegal device, and I make the device the organizing analytic of this book.

DEVICES

Each chapter in this book centers on one of four devices—formula, index, list, and pact. All of these devices are inscribed in larger processes of water price setting, legal reform, or the promotion of care for water. They are also pieces in even larger trajectories of globalization, the financialization of water, the judicialization of politics, and even the nationalist, neoliberal redefinition of the public sphere we are witnessing. We could begin analyzing these devices by asking questions about those macrohistorical processes, bounding their significance to a specific role in those larger happenings. That approach would turn each device into a token of larger political and economic contexts, namely the history of welfarism in Costa Rica or oppressive patron–client relations in northeastern Brazil. In this book I want to sidestep that token relationship and take a different approach. I will remain close to the morphology of the device, attending to its variations and textures, to its crevasses and revelations, in order to capture the power of seemingly minor technopolitical decisions to shape the abstraction and rematerialization of water. By attending to the form and liveliness of these devices, we gain a different analytic entry point to see how people mobilize history, knowledge, affect, and ethics in their daily professional and political lives. I take this approach because while we search for new macroschemas to adequately address ongoing struggles over things as basic as water, many fundamental ethical questions of our time are being answered quietly, almost inadvertently, through devices like the ones I study. I believe that better understanding their intricate details allows us to imagine new forms of technopolitical mobilization; these devices can open space for new future histories.

A device is a highly effective instrument for organizing and channeling technopolitical work.[3] It is a technical instrument that merges practices and desires with long-standing assumptions about sociality that have been embedded in legal, economic, and other technical vocabularies and institutions. A device is a structured space for improvisation; it is embodied in the actions of specific persons, but it is also a braiding of long histories of economic, legal, and political systems. In my conceptualization, a device both affirms and destabilizes social categories and institutions, while providing a way to identify the particular practices, offices, computer files, and conversations whereby that material-semiotic labor is performed.

Given this capaciousness, I think of a device as an intense node of temporalities and passions, a combination of diverse technical inheritances (the history of ideas) that open the possibility for other possibilities.[4] A device opens space for technical improvisation even if it is often described by highlighting its fixity, as if its components were already predetermined. People constantly engage these devices through tweaks and hacks that make the technical traditions that seem to be already ordained more flexible and open than they appear. That simultaneous fixity and openness gives a device its capacity to affirm and destabilize social categories and institutions. But, as I mentioned above, it also gives the device its concreteness, allowing us to identify the particular subjects, practices, and locations where we can study them ethnographically.

Although producing diverse constellations and forms of water, the devices I analyze in this book are deceptively humble. In our conversations and work together, my interlocutors were not shy about reminding me that they were fully aware of the precarious nature of their devices, yet, at the same time, they insisted that despite such precariousness, their work consisted of pushing those tools to their limits and getting them to do as much work as possible. Inaction was not an option. While having an unassuming appearance, these devices have the capacity to effect important differences. After all, as the history of Christianity shows, an iota of difference, a barely perceptible divergence, can divide nations, religions, and the histories of whole continents.[5] Some of the devices I study in this book emerge from a particular body of knowledge, as in a mathematical formula; others result from people's lived experience, as in the creation of a pact to care for water. And while formulas, indices, lists, and pacts are portable and can travel across geographic locations, the results of their activation are never homogeneous. Each time a device is used, its outcomes vary in small and large ways.

I encountered these devices in the manner that other anthropologists encountered necklaces and arm-shells when they asked people in the Trobriands about their valuables (Malinowski 1920; Weiner 1985, 1992). When I asked my interlocutors about the future of water, they explained the need to differentiate a human right from a commodity and immediately referred to the devices they were using to achieve that goal. What I originally imagined were going to be discussions about moral values and ethical futures quickly shifted to explications of the tasks of calculating a formula, designing an index, delineating a list, and securing a pact. As it turns out, these devices are the means by which people clarify moral preferences and enact temporal assumptions about the "goings-on" of life. I imagine these devices as something akin to a gadget, a small thing with aptitudes to crystallize regimes of technopolitical value and relationality. These seemingly small devices help people carve out a sense of what a good common life could be, though they may also often undo that very same sense.

Understood in this way, the devices in this book possess capacities similar to those of complex words (Empson 1977). They create space for the play of ideas and their "histories, transformations and divergences," while exerting pressure on that creativity to stay within particular parameters (Swaab 2012: 272; Williams 1977). These devices create conditions that make some decisions predictable, as when an inflation index is the go-to resource to adjust the price of water, while in other cases they compel people to lift the rug to see what things have been "swept under" it in the rush to deal with pressing problems, as when people unwittingly generate a taxonomic list to legally define what water is.

In this conceptualization of a device I attend to its semiotic charge as developed in linguistics when we talk about a stylistic device or device of speech. I also attend to its technicality as investigated by science and technology scholars who remind us to ask questions about epistemic histories and material configurations. And, I also pay attention to the political capacities of a device as mapped through governance projects that depend on disciplinary associations of knowledge/power as diagnosed by Michel Foucault. But these theoretical markers are labels that I assign to them a posteriori, after having encountered them in the world. So, while I offer these ideas as guideposts for the reader, I am more interested in developing the potential of the device as an ethnographic category. If these devices are practices in the world, they also affect the world by creating new categories. I want to suggest that devices are not only good things to think with, but also good thoughts to act with—for ethnographer and interlocutor alike. They help us create concepts to make sense of the world, and they make worlds in relation to concepts.

Consider, for instance, the act of haphazardly producing a working list of types of water to be covered by a constitutional reform to recognize water as a public good and human right in Costa Rica (see chapter 3, "List"). Such a device, the working list, has a dual power. On the one hand, it reveals what seems implicitly reasonable: the types of water that should be considered a public good to guarantee universal access. On the other hand, the items on the list open an opportunity to propose a different arrangement, to come up with an unconventional answer for the simple question of why things are the way they are. What if, say, rainwater were included in the list of public goods? How would that change the distribution of matter, entitlements, and costs? How might that alter the very idea of a human right? In Costa Rica that list has occupied more than fifteen years of congressional sessions devoted to the discussion of constitutional reforms. While taken seriously by some and used by others as an excuse to ridicule the idea of a human right to water, the list and the procedure that made it possible have functioned as a wedge, carving out space for discussions of the strategic, the self-evident, and the nonsensical. The list's capacity to absorb the energy of those participating in its construction has turned it into a symbol of effervescent political polarization that has almost exhausted the will of those promoting the human right to water.

What follows, then, is an examination of how categories, practices, and devices animate social worlds. I have put together a collection of four devices, three from Costa Rica—formula, index, list; and one from Brazil—pact. The three Costa Rican devices are all highly technical instruments that required a lot of effort to make sense of. I not only had to follow practices that are not readily available for observation, many of which included people sitting at their desks; I also had to familiarize myself with economic and legal technical languages, and with the rules of congressional procedure. All three of these devices are critical passage points in bureaucratized processes. It is not surprising that studying the creation of differences in Costa Rica takes this form. Today, environmental politics and really most mobilizations to address collective life in relation to the state take a fragmentary and piecemeal approach. There is no sense that all-encompassing change is possible in the country. Rather, there is a feeling of things being stuck the way they are. If by a stroke of luck transformations are brought about, they are piecemeal, only one small step at a time.

When my research in Brazil began, it was striking to me how different the political mood and sense of possibility was in comparison to Costa Rica. In my first days of fieldwork, particularly in early visits to the state of Ceará's regulatory agency, I encountered technopolitical devices similar to the ones I followed in Costa Rica. Lula da Silva, Brazil's leftist president, was in power and in Ceará a conservative governor was in his last term. The whole country was wrapped in a mood of profound transformation. There was an intoxicating sense of openness. A year or so later, after I arrived in Ceará again, I found something else was happening besides what I had noticed in the regulatory agency—there was a process that was touching, in one way or another, almost all the water activists and experts that I knew. That process was the creation of the Water Pact (WP), an ambitious statewide effort to promote care for water among all of Ceará's citizens. To me, it was notable how the pact was predicated upon the possibility of massive change, of transforming society as a whole. The rationale and techniques the pact organizers relied upon were geared toward "large-scale" visions, ways to aggregate the political will of "all of society." While the devices in Costa Rica focused on more narrow issues, the pact was an attempt to effect larger-scale change. I switched my focus and made the pact the focus of my fieldwork.

That is how this collection of four devices, one from Brazil and three from Costa Rica, came into being. I have preserved the distinct tones of each device throughout my writing, in part to keep in mind that there is nothing set in stone about the form, scale, intent, or motivation of a device. All of those are questions that have to be ethnographically elucidated. Furthermore, I have also tried to preserve their asymmetric scales in order to convey the sense of fragmentation, lack of closure, and comprehensiveness within which my interlocutors conduct their work and attempt to change their worlds. Yet, all of the devices I have followed are experienced as one possibility among many. My interlocutors commit to that possibility, accepting its legacies and hoping for its potential to be achieved, but they are aware that with their selection they have no monopoly over the future. The devices that they use to help organize their technopolitical labor are, most of all, just one of many possibilities.

By analyzing these devices, along with the intellectual and affective passions they ignite, I want also to mirror the temporality of social life as it is experienced by my collaborators: amid unknowns and without the certainty of hindsight. Theirs is a world in process, experienced from within the instabilities of the present. This temporal orientation allows me to keep in sight how my interlocutors selectively activate certain histories and how docile they are in the face of dominant stories of the past (Bergson 2002; Chakrabarty 2000). This temporal orientation also keeps us attuned to the contradiction and trepidation inherent in all technical acts. I will argue that this temporal orientation is necessary if we are to carefully interrogate the contradictory possibilities of all technical processes, and even more so at a time when water's tendency to change material form disorients our inherited environmental, political, and economic categories. Under these temporal conditions, neither dreams of intimate access to people's worlds nor the promise of distant structural diagnoses of historical developments can do the necessary analytic work. We need alternatives to this prevalent analytic dyad. I propose using the device as an one such analytic alternative.

WATER, WORD, AND MATTER

Because of its universal multiplicity and predisposition to vary its material and abstracted forms, water often confounds any attempt at fixity (Helmreich 2015; Linton 2010). Water's significance for the sustenance of life makes its symbolic meaning multiple (Strang 2006). But its material form is also multiple, destabilizing any schematic rendering of what a water body is. For one, water's defining trait is its tendency toward the formless, its obsession with gravity, its material inclination to change. The French modernist poet Francis Ponge describes this condition by saying that "water collapses all the time, constantly sacrifices all form, tends only to humble itself, flattens itself onto ground" (Ponge and Brombert 1972: 50). Alternatively, we could say that it is not its lack of form but water's magnificent capacity to take a huge variety of forms, the infinite metamorphoses it is capable of—spouts, streams, pools, fast or slow flowing, whipped into turbulence, pulled by the moon, soaking things, and finding its level at rest—that creates the challenge of finding ways to engage its significance for life (Marilyn Strathern, personal communication, April 6, 2018). This characteristic tendency toward morphological reinvention (Ballestero 2019)—water's proclivity to flow, freeze, and vaporize—confounds the institutional and organizational protocols we use for its scientific exploration and political organization.

This kind of unstable relation between knowledge and material bodies is not unfamiliar to us. Feminist scholars of science and technology studies (STS) have taught us to think about it in terms of the material-semiotic and to consider how corporeality is, at once, a force that shapes knowledge and a substance that is shaped by it.[6] Bodies, human or watery, are not preexisting entities, nor are they purely ideological. They "are effected in the interactions among material-semiotic actors, human and not" (Haraway 1992: 298). Matter, as concept and thing, "is itself culturally and historically specific and, as such, contested terrain" (Willey 2016: 3).

Feminist STS scholarship has helped us see how the types of knowledge and tools doing the morphological work of defining material bodies are scientific. But we sometimes forget that they are also legal and economic and that all of these forms of knowing can work together to specify what water is. Regimes of exchange, for instance, accord certain materials with some values and properties but not others. The water in a bottle bought at a grocery store is a different substance from the water poured into a bottle from a well on public lands. It looks different, and often tastes different (Spackman and Burlingame 2018). Take the case of Ceará, where people in the rural areas install fences made of wire and dry wooden branches to create property lines. These fences often cut across water bodies, small or large ponds. When the dry season sets in, most water bodies dry out slowly, revealing to landowners that their carefully placed fences hang in the air, clinging to the shores of a pond that was, might again be, but has disappeared. These hanging fences now cut the air in two, as if mocking the figure of property, at once showing the violence and absolute fragility of the separations they produce. These appearing and disappearing water bodies, and the fences that cut them through, not only shape everyday household and agricultural routines by demarcating where water is accessible and for whom, they also reveal the seasonal specificities of legal and economic relations forged around the presumed stability of a property regime that allows landowners to sell water for profit, commodifying its life-granting properties. These cyclical transformations of sociomaterial forms marked by hanging fences capriciously activate and mute obligations, the movement of cattle, amity and dispute between neighbors, political relations of debt, and the power of the state to move water in cases of emergency. Property lines attempt to define water morphologically.

As this example reminds us, regimes of knowledge (science), obligation (law), and exchange (economy) constantly shape what we count as material. They determine the matter we enroll into relations of credit and debt, into the very definition of what a basic human need is, and into the categorization of nature as such. The point I wish to emphasize for us to keep in mind throughout this book is that in the making of matter, not only scientific word and measurement are entangled with substance (Barad 2003). Legal and economic forms of knowing also perform those kinds of material configurations and, more often than not, they do so from a distance.

From this point of view, apprehending water materially cannot be limited to a supposedly stable form of H_2O from which we can infer cultural or political consequences of its presence or absence. Thinking about the materiality of water entails querying, first of all, what its corporeality might be, how something becomes a water body in a particular time and place, and how that body is always a technopolitical entity. It entails attending to how its contingent presence is brought about by much more than our scientific capacity to comprehend bonds between hydrogen and oxygen (Sawyer 2017).[7] As I will argue, we need to remain attentive to the capacity of technolegal devices to implode the supposed material certainty of the molecular. We need to trace water itself beyond pipes, dams, rivers, and oceans. Thus, in what follows, I focus less on watery scenes, fluid locations, and aquatic environments, and instead focus intentionally on water elsewhere, in places where we might not usually explore its material politics.

Diagnosing the existence of such entanglements between legal, economic, and scientific word and matter is not enough, though. Stopping at this diagnosis would leave us at the point where we should just be starting. One of my central interests is to think about what comes after material-semiotic entanglements have been diagnosed. What do people do when entanglements are part and parcel of their sense of the world? As I show, one of the things people do is to reflexively separate that which they encounter and understand as already knotted. They try to undo the entanglements they encounter. This returns us to the issue of how people create bifurcations amid the intense relationality of word and matter. The devices I study in this book help people transform fusions into momentary separations; they allow people to create separations to cut and redirect relations so that bifurcations can be effected. Furthermore, it is through their devices that people channel their efforts to theorize and organize the ethical responsibilities that emerge from the ontological surgeries they perform (Jasanoff 2011; Valverde 2009). Creating separations is sometimes the only ethical way out.

HUMAN RIGHTS, COMMODITIES, AND THE SPACE BETWEEN

During the first decade of the twenty-first century, the international establishment saw the idea that water should be a human right as contentious. All sorts of interpretations circulated about its implications. A water policy expert from the United Kingdom whom I met at the Stockholm Water Week in 2009 told me emphatically, "The problem is that those who want water to be a human right don't understand that somebody needs to pay to bring it to people's houses. They want water to be free. And that is just unviable." He was among the progressive proponents of universal access, yet he feared that such universalism could be made so profound that it would cause the financial collapse of the water sector. His worry was universal, totalizing. I was surprised by his argument, in part because none of the Latin American activists with whom I had worked for years had ever suggested that water should be completely free. They had a nuanced understanding of the financial and physical challenges of moving liquids across vast open landscapes or packed urban conglomerates—the difficulties of controlling pressure, flow, and leakage, and the monitoring toil of keeping water molecules as pure as possible. Yet the message that "activists" wanted water to be free carried a lot of weight and was mobilized by many to discredit the aspirations of those demanding more democratic access (see also Schmidt 2017).

By 2015, only six years after my conversation at the Stockholm Water Week, the terms of the debate had changed drastically. The international establishment seemed much more accepting of using human rights language to make the politics of water speakable. Perhaps this was due to the fact that in 2010 the UN General Assembly officially recognized the existence of a human right to water and sanitation through resolution 64/292, which cited multiple preceding declarations, events, and projects showing that this was a decision long in the making (see figure I.3). Or maybe it was because eleven Latin American countries, among others around the world, had modified their constitutions or passed new water laws to formally rec-

United Nations A/RES/64/292

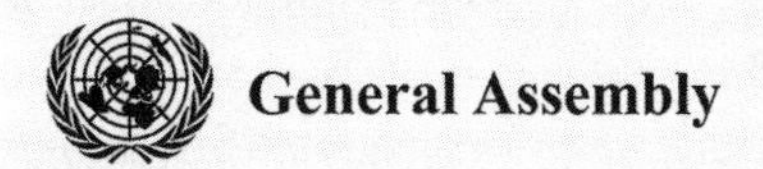

General Assembly

Distr.: General
3 August 2010

Sixty-fourth session
Agenda item 48

Resolution adopted by the General Assembly on 28 July 2010

[*without reference to a Main Committee (A/64/L.63/Rev.1 and Add.1)*]

64/292. The human right to water and sanitation

The General Assembly,

Recalling its resolutions 54/175 of 17 December 1999 on the right to development, 55/196 of 20 December 2000, by which it proclaimed 2003 the International Year of Freshwater, 58/217 of 23 December 2003, by which it proclaimed the International Decade for Action, "Water for Life", 2005–2015, 59/228 of 22 December 2004, 61/192 of 20 December 2006, by which it proclaimed 2008 the International Year of Sanitation, and 64/198 of 21 December 2009 regarding the midterm comprehensive review of the implementation of the International Decade for Action, "Water for Life"; Agenda 21 of June 1992;[1] the Habitat Agenda of 1996;[2] the Mar del Plata Action Plan of 1977 adopted by the United Nations Water Conference;[3] and the Rio Declaration on Environment and Development of June 1992,[4]

Recalling also the Universal Declaration of Human Rights,[5] the International Covenant on Economic, Social and Cultural Rights,[6] the International Covenant on Civil and Political Rights,[6] the International Convention on the Elimination of All Forms of Racial Discrimination,[7] the Convention on the Elimination of All Forms of Discrimination against Women,[8] the Convention on the Rights of the Child,[9] the

Figure I.3. United Nations General Assembly resolution recognizing the human right to water and its international law precedents.

ognize the human right to water (Mora Portuguez and Dubois Cisneros 2015). News about the passing of each law or constitutional reform circulated through the activist and water policy circles I was part of as evidence of a better future that would soon arrive. Human rights offered something of a counterweight to both the privatizing efforts that had swept the region during the 1990s and the hype for public–private partnerships to modernize water management of the 2000s.

A YouTube video of Nestlé's CEO, watched by thousands globally, provides more evidence of how quickly things had changed. The video showed a 2005 interview conducted in German with, depending on the version of the video you saw, a slightly different translation of the CEO's words. In all versions, however, he claimed that water should be managed through markets, like any other commodity, and should not be treated as a special right.

A few years later, Nestlé's CEO reversed his position. Explaining that his former comments were taken out of context, he began presenting himself in venues such as the World Economic Forum as an avid supporter of the human right to water. Reversals like this have led people to regard human rights as weak anticapitalist tools. If, during the 1990s and early 2000s, activists and some water policy experts had trust in what the recognition of the human right to water could accomplish, today, their commitment is more nuanced. The boundary between a human right and a commodity is blurrier than ever. Nevertheless, they continue to push for the human right to water but with much more modest expectations.

The widespread worry over the commodification of water among the activists and experts I worked with is far from unwarranted, despite the slowing down of the privatizing fad of the 1990s. In the early 2000s, for instance, *Fortune* magazine reported that only 5 percent of the global water industry was in private hands, leaving a great potential for untapped business opportunities for the expansion of private enterprise. Global banks such as HSBC advertised their services by posing questions about the financial value of water, narrowing its existence to a luxury or a commodity (see figure I.4). Supplying water to people and industry was at the beginning of the twenty-first century a $400 billion-per-year business, equivalent to 40 percent of the oil sector (Tully 2000). More recently, RobecoSAM (2015), a financial company based in Switzerland that focuses on environmental and sustainability financial investments, considered water "the market of the future" and described its current financial landscape in the following terms: "Recent estimates put the size of the global water market at about USD591 billion in 2014. This includes USD203 billion from municipal capital expenditure, USD317 billion from municipal operating expenditure, USD1 billion from industrial capital expenditure, USD 37 billion from industrial operating expenditure, USD12 billion from point of use treatment and USD3.7 billion from irrigation. Market opportunities related to the water sector are expected to reach USD1 trillion by 2025" (20).

It is striking that of those US$591 billion that they calculated in 2015, US$500 billion are invested, allocated, or directly managed by municipal or public entities. While environmental analyses emphasize that most of the world's water, between 70 and 85 percent, is used for irrigation, the overwhelming majority of the "market share" RobecoSAM is interested in is public or municipal provision for human consumption and industrial use. In other words, the distribution and structure of the financial universe

Figure I.4. Banking ad using water to establish a universe with two possibilities: commodities and luxuries.

does not match the hydraulic universe. Tracing where most H_2O flows to and from does not necessarily take us to the areas where most financial attention is put. This means that the way water prices are set, the legal categories countries adopt, and the quantity and types of subjects they recognize as users entitled to the human right to water are all decisions that directly shape desires for financial returns, international investments, and the global relation of water to capitalist wealth and profits. Financialization affects the routes, pressures, and qualities of the flow of water as well as the global accumulation and distribution of "market" opportunities to increase returns.

In schematic terms, commodification is the process of making an object commensurable with other objects so that its exchange is possible within market-like formations. In Marx's famous formulation, commodification turns qualities into quantities through a variety of technical and magical means that make things that are intrinsically different appear, if only temporarily, as equivalent. As things are commodified, boundaries are rearranged, social relations and significations are transformed (Helgason and Pálsson 1997: 465), and relations between people and things take the form of relations between things (Gregory 1982; Mauss 1967). Of course, this is not a mechanical or smooth process. Water is, to use Radin's (1996) words, a "contested commodity" that poses cultural and affective difficulties for its complete commodification because it remains embedded in different, unstable meanings, and for that reason is always gesturing toward the possibility of forestalling the equivalence on which its commodified exchange depends. For Polanyi (1957), water is a "fictitious commodity" because no labor has transformed its essence and hence it fits better in the realm of "society" and not in the realm of the capitalist economy. But also in this sense, the character of water is slippery. Within a single community, people can think of water as sacred, store it, reject its exchange, or pass it on as a gift of nature. They can also pay a water bill at the end of the month, buy bottled water from a store, and pay a neighbor to connect to their line. Even if at some point water is commodified, its social life entails a moment of decommodification to be ingested, shared, or bathed in. The economic biography of water is always a rich series of transformations of its value form (Appadurai 1986; Kopytoff 1986).

In order to understand such mutability and the different obligation regimes associated with it, anthropological analyses of commodified forms of exchange rely on a contrast with gift economies to make the particularities of each clear. Building on Marcel Mauss's (1967) foundational text, anthropologists conceptualize gifts as singularities, things whose value is not assessed through universal equivalences but brought about in singular regimes of exchange that differ in temporalities and rules from those organized around commodities (Munn 1986; Strathern 1988; Weiner 1985). If commodities facilitate the smooth exchange of value given a predetermined medium of equivalence—money—gifts enact exchange via the intensification of particularities and their variations according to context, social status, gender, and history.

Put in this way, the contrast between gifts and commodities seems much more stark than it really is. Ethnographic examinations of these modalities of exchange since Mauss have shown that gifts and commodities are not alternative regimes but idealized types of sociomaterial relations that coexist in all kinds of creative combinations.[8] The rich economic biography of water is also evidence of this. Nevertheless, it is not the ethnographic record that interests me here. Instead, I want to consider what happens if we dispense with the analytic prevalence of the gift–commodity dyad—and more to the point, what happens if we think instead of the commodity–right relation.

While marginal in comparison to the gift/commodity opposition, the relation between liberal rights and commodities has not been absent from cultural analysis. If gifts and commodities have been imagined as an opposition, rights and commodities have been conceptualized homologically—that is, as operating on similar principles and structures. Isaac Balbus's (1977) classic work, for instance, offers a powerful theorization of these two figures. Building on the work of the Marxist legal scholar Eugeny Pashukanis (1980), Balbus shows that the law operates under the same assumption of equivalence that allows commodities to exist. If commodification is the process of turning a use value into an exchange value that can be expressed in a common medium, fundamental rights work in the same way. Individual citizens with all their particularities and idiosyncrasies are made commensurable to each other through their fundamental entitlements as rights-bearing subjects. Fundamental rights perform the magic of equivalence by erasing the marks that birth, gender, social rank, education, and political affiliation leave on our embodied experience. This commensuration makes possible representational democracy and market ideologies alike, as they similarly depend on purported equivalencies (Balibar 2004; Collier, Maurer, and Suarez-Navaz 1997).

I return to these classic texts because their insights create a point of convergence between scholarly works and my collaborators' own theorizations of their water conundrums. On a number of occasions, as we analyzed the legacies structuring water inequalities in Brazil and Costa Rica, my collaborators and I engaged in extended conversations on whether and how Marxist theory helped make sense of this homology. The question for them, and for me, has been the following: Once you are aware that commodities and human rights are, in a sense, the same thing—entities that share a structural form—what can you do about it? This homology between human rights and commodities poses crucial questions about how to imagine the possibility of making a difference—and changing the world—through these figures. Thus, it is not surprising that for the people among whom I worked, as well as for others for whom rights and commodities are important categories for the organization of social life, the act of creating distinctions is a critical one.

AS IT IS, BUT DIFFERENTLY

On one occasion, after sharing my initial findings with colleagues at Ceará's water agency, a geographer in the audience posed a piercing question. He asked me whether I had considered how Karl Marx had analyzed the issue I was interested in. He reminded me of Marx's ideas about commodification and class struggle. At the moment, I was unsure about how to answer his question but I took his observation to heart. I later returned to Marx's work, particularly to a lively passage in volume 1 of *Capital* (1976) where he explicitly addresses the problem of the relation between rights and commodities and the question of singular or multiple worlds. Marx notes how seeing liberal capitalist society from two different perspectives reveals two coexisting, yet distinct, spheres of action. From one point of view, an observer can see the order of rights where an individual can legally express her willful existence as a subject. That, Marx tells us, is "[t]he sphere of circulation or commodity exchange . . . the very Eden of the innate rights of Man" (280). But when the observer leaves this sphere, what seemed equal is revealed as asymmetrical, and the "physiognomy of our *dramatis personae*" changes drastically (280). The money owner becomes the capitalist, and the one who sells labor power becomes the worker. The relation between them is now asymmetrical and the equality that liberalism promises is revealed as only illusion.

With Marx's assistance, activists and experts in Costa Rica and Brazil take on the challenge of puncturing this illusory relation between different spheres of action and the homology between human rights and commodities. They know all too well that rights and commodities are not as radically different as they once seemed. With this recognition they try to deal with the differences between these spheres of action while inhabiting a world that they do not have the luxury of leaving, a world that has no radical alternative, an Otherwise, readily available. Thus, instead of attempting radical alternatives, they search for differences through a practice of proximation as the only way of "noticing particular moments . . . where interesting forms of friction or tension emerge" (Gad and Winthereik n.d.: 3). In this world, making a difference requires getting closer to, not distancing oneself from, what is already in place.

Many anthropologists and activists have considered the question of one or many worlds. At locations such as the World Social Forum, activists entertain the question of multiple worlds by saying that *otro mundo es posible* (another world is possible). This phrase signals a commitment to a politics that assumes that the world can be organized differently. But I want to call our attention to the fact that this imaginary of other worlds, held by activists and ontologists alike, depends on a sense that combines multiplicity and exteriority in a quintessentially modern form. During premodern times of theological social order, the world was one way, and that singularity was preordained by a superior entity, God. There was no outside or exteriority to that order. In modernity, we understand that it is up to us and to our social institutions to structure the worlds we live in. It is only in this world that the very possibility of being otherwise is conceivable. While one could assume that the technical worlds and discussions that I analyze here belong to a modern understanding of a world that is to be molded to one's desires, I found the opposite. In my collaborators' technical worlds, multiplicity and exteriority are absent. For the people with whom I worked, the world is one and it can only be rearranged using existing resources and ideas. The world that is possible is the one that is done and undone in front of them. In this world, ontological difference can only be rendered as opinion, not fact. Thus, rather than seeking a new perspective from which to access a different world, they mobilize to create a difference in the world that is. The picture people see is singular, a sort of legal and economic mononaturalism (Descola 2013) that challenges the very foundation of the modern liberal order: the belief that that there is an outside waiting to be inhabited.

In such conditions, when the world in front of us is the only actionable one, the issue is how to make a difference emerge, how to create the conditions to make difference visible in a world where the precise terms of a bifurcation are never stable and what currently is seems to leave no space for things being Otherwise. The epistemic and ontological problem here is how to create enough separation so that distinctions are possible even if the world seems to preclude any permanent differentiation. Put another way, the people working to have water recognized as a human right imagine a difference that makes a difference without resorting to radical alterity to create a contrast in perspectives.

I want to suggest that this practice of creating a difference without resorting to radical difference or the Otherwise is a project that entails committing to the world *as it is, but differently*. That is, it is a commitment that entails a mode of purposeful engagement that unfolds without presuming that one's desires or systematic interventions have the power to produce a radical difference in the course of history, yet recognizing that within that apparent immutability there is open space for play. This orientation requires a form of inhabiting the world that is not pluralistic, not organized around a multiplicity of worlds that can be placed side by side for an observer to choose which to step outside of and which to enter. After all, that dream of stepping outside of what is—of being illiberal or aliberal—is one of the most fundamental assumptions of modern liberalism. Instead, in this world difference has to be worked from within, as a labor of recuperating that which has been discarded as inconsequential. This is a commitment to the world as it is while trusting that there might be a chance to qualify it differently and, by doing so, to inhabit it more purposefully. Understanding how people act tactically in that world requires us to hold in abeyance our anthropological assumptions about difference as self-evident multiplicity.[9] Another implication of this is that rather than presuming that difference is the "natural" condition of social worlds, we begin to see differentiation as one possibility among many and, for that reason, one that necessitates considerable epistemic and ontological labor to be accomplished. And finally, engaging the world as it is, but differently leads to a peculiar relation to the future.

FUTURE HISTORIES VS. HISTORIES OF THE FUTURE

So far, I have referred to Costa Rica and Brazil as locations where I have conducted fieldwork. I have also mentioned the World Water Forum, that triennial event that attracts world water elites, as a space I shared with my interlocutors. But I have not settled on a single geographic site as the location for the stories that you will find in this book (Gupta and Ferguson 1997; Marcus 2006). One reason for this is that this book is better located in time than in space. It is ethnographically grounded in the device, with its impetus to "improve" futures. Here, I am not suggesting a dichotomous separation between time and space. Rather, I am engaging in an exercise of emphasis, making the choice to put more pressure on one analytic thread, the device and its temporality, to find out what interesting insights it can generate.

This emphasis on the device as an intense node of temporality is crucial to understanding its character as simultaneously precarious and hopeful. In my conceptualization, the device opens up a conditional temporality where encounters between the material world, the body, tools, ideas, and representations (Bear 2014: 20) shape collective senses of accountability and plausibility (Greenhouse 1996, 2014). Those encounters occur in community aqueducts, bureaucratic cubicles, signature collection campaigns, international meetings, and moving vehicles promoting citizen participation in water management. My ethnography will convey those specificities. Yet the everyday work occurring in those locations is also connected to medieval economic history and notions of profits (chapter 1, "Formula"), inflation rates and the collection of household objects (chapter 2, "Index"), political communities that challenge Leviathan's singularity (chapter 4, "Pact"), and attempts to draw the material borders of water bodies (chapter 3, "List"). How can one keep all of those connections in sight? This book attempts to do so by thinking about nonlinear future histories of water. It attempts to show how people relate to future histories without falling into predictive modes. It shows instances where engaging the future does not necessitate having an image of how that future looks.

While analyses of the future often emphasize its openness and unpredictability, the future is anything but empty. We are surrounded by, or have the habit of looking for, proleptic images. Even if we know that those images are not certain, we still rely on their contents. This is the paradox of modern futurity: while we are taught to believe that the future is unpredictable, we live in a world saturated with future-consciousness (Rosenberg and Harding 2005: 6). A history of the future, Rosenberg and Harding note, shows how futures—as (meta)narratives that foresee, predict, imagine, divine, prognosticate, or promise—encounter people's everyday lives (9).

Through the devices that I study in this book, I engage the future in another way: I move from histories of the future to conceptualizing future histories. By inverting these two concepts I want to tap into the nonimaginable dimension of the future. But this nonimaginable future is not unimaginable because it is too traumatic or extreme. Rather, it is unimaginable because of its unpredictability. There are no metanarratives to connect it to people's everyday lives. The uncertainty is too deep here and is the result of an awareness of the interminable practices, material processes, imaginaries, and mere coincidences that ultimately shape the yet-to-come. For me, it is not surprising to find this mode of addressing futures within bureaucratic-like spaces (Mathur 2016). This is not a modern future. It is not foreseeable, predictable, or imaginable. And yet, despite its "unimaginability," it is engaged through the density of everyday action. Thus, my conceptualization of a future history signals happenings that will be recognizable as meaningful only from the future; only by looking back will what counts as the history of an event be recognizable. The devices I analyze have the potential of becoming that future history or, at least, of creating its preconditions.

Anthropology's record of thinking about the relation between sociality and time has produced rich analyses of people's orientation toward the past—from evolutionary theories to recent and personal histories (Munn 1992). Recognizing the future as a "cultural fact" (Appadurai 2013: 285), anthropologists have shown the medium term can be evacuated from collective preoccupation (Guyer 2007), how the future can be ossified as a site of nuclear disaster (Masco 2014), how "anthropocenic" ends of the world are diagnosed (Cohen 2016), and even how the future has operated as the very ground of anthropological analyses (Ringel 2016).

Feminist thinkers, on their part, have also long reflected on the future. I am interested here in the work of feminist scholars who have invited us to think about how the future can be "conceptualized in different terms" (Grosz 2002: 13). I take this invitation to search for alternative conceptualizations as a call to replace the quest for what the future looks like with the question of what counts as the future in the first place. Within this frame, we can move from a search for narratives of beginnings and ends, in the form of images of rebirth or apocalypse (Wiegman 2000), to a focus on questions of duration and of the political possibilities of the in-between. In that space, the question of affirming the worlds that we want to inhabit acquires a more intimate scale, challenging the comfort of critique, if critique is defined as a distant diagnostic of negativity (Braidotti 2008). Rather than undoing worlds or focusing on documenting their lacks, this feminist future poses questions about the ways in which worlds are remade in what we understand as the goings on of the present.

Other academic and professional disciplines—such as neoclassical economics, statistics, and more recently environmental and earth sciences—constantly attempt to produce the future by relying on visions of the world to come. Using sophisticated techniques of calculation, modeling, and planning, and relying heavily on computerized procedures that process large quantities of data, these disciplines routinely produce image-like iterations of how the future might look (Mathews and Barnes 2016). Inscribed in the methods by which those visions are put together we can find assumptions of what is possible and what is plausible. Those assumptions about what counts as relevant information for future making result in a picture of how things could be (or not). They result in a future that is seen in the body of an inflation percentage, a number of people with access to clean water, a situation where all water is managed by privatized utilities.

For some social commentators, the devices in this book might seem tools to make exactly those kinds of visions of the future concrete. But I will argue otherwise. I will show how, given their openness, these devices allow people to not engage the future as if it were an exhibition, a display you could step into, or even as a narrative figure. My interlocutors do not use the liveliness of their devices to produce a utopic, dystopic, or merely unremarkable image of the yet to come. This refusal to treat the future as an image is not capricious. It is intrinsic to the work of creating bifurcations between terms when you know those bifurcations are inherently temporary, and when you are aware that any difference created in the present is unstable and contradictory, despite the potentially brutal effects it might have. Rather than talk about the future they want to see come about, they speak about responsibility, principles, and shortcomings in their technical acts. This is how they create a future history, not by talking about what that future looks like, but rather by acting in the present with all its constraints and limitations.

When people activate the future and their devices in this way, they stretch themselves between different moments in time simultaneously. They activate the legacies inscribed in their tools, they mobilize what they recognize as the present, and they project both into a sense of the future as something one is responsible for in the here and now. In other words, they create a temporality that folds linear order onto itself. Futures and pasts are brought into the present, turning now into something more than what we think it currently is.[10] Instead of a chronological unit, that moment is a simultaneity full of conditionals, dependencies, and uncertainties that cannot be compressed into an image. If that moment is turned into an image, it has been turned into something else. It has become a predictive, and hence incomplete, vision.

Thus, instead of relying on a fixed image of the future, my collaborators deal with that simultaneity by thinking and acting from the future anterior, that upcoming moment that "is not calculable from what we know, [because it is] a future that surprises" (Fortun 2012: 449). In this temporal orientation, the devices people use and the multiple bifurcations they negotiate are processed with the expectation that they might work as preconditions that pave the way for something that is different from what is. Yet, despite their technicality, my interlocutors cannot know exactly what the preconditions they help create might accomplish in the future. This future anterior is actualized in those practices of the present that embrace the future's impossible calculability, without relegating it into the unthinkable or into a realm of ideas that cannot be acted upon. In this folded temporality, people act by setting up "structures and obligations of the future" (Fortun 2012: 449), despite the difficulties they have with producing any specific image of what that future might look like.

Analyzing that temporality complicates our ethnographic confidence in the historical as a fait accompli waiting to be described. Here, ethnographic analysis cannot be limited to a narration of events that have already occurred as if their significance lay in their pastness. Nor can analysis be guided by the temporality of nonevents, those everyday actions that are illegible and insignificant for dominant collective schemas. An ethnographic analysis of the future anterior traces a three-way temporality: the possibility inscribed in future differences, the past legacies shaping accountabilities, and the present opportunities mobilized to foster unanticipated plausibilities. In this temporal mode, people act to set up structures and obligations for the yet-to-come, despite their inability to visualize that future precisely. This book shows how people engage in that work. It is not that they pose the question of temporality as a topic to be discussed; rather, I show how they produce differentiations when their everyday work is already marked by a particular sense of the yet-to-come.

An ethnography written from this temporal orientation can leave one's desire for completion unsatisfied. The future anterior is not built on nonevents, those happenings that go completely unnoticed or unrecognized by dominant forms of reason (De la Cadena 2015: 145–48). Rather, narrating difference in the future anterior depends on *quasi-events*, things that are not privileged by a sense of full existence but instead unfold without "quite achiev[ing] the status of having occurred" (Povinelli 2011: 13). The devices this book examines are quasi-events themselves: lists put together without ever becoming law, percentages of surplus never increased, promises aggregated without having their fulfillment verified. Discarding those occurrences on the basis of their lack of "effects," where effects are predetermined by what we can recognize in the present, would close off our access to possible futurities. It would keep us tied to the familiarity of the predictable. This means that writing ethnography from the uncertainties and conditionals of the future anterior is writing what might become a future history, something that from the future might provide insights into how what currently is has come into being. This is why I want to argue that dwelling in what in the present seems to be ineffectual is a worthy analytic endeavor.

In an effort to attend to that temporality, I explore the collection of devices I have curated by spending time within folds and tweaks so that we can recognize the efforts people make to set up future differences, or at least to create their preconditions, even if we cannot round off their stories with an end point. This approach allows us to create an "opportunity to arouse a slightly different awareness" not only about "the problems and situations mobilizing us" (Stengers 2005: 994), but also about the ways people confront those problems. This attentiveness also has the peculiar effect of making certain bifurcations more perceptible, turning significant that which otherwise may seem irrelevant. And finally, this approach also affords us some time to wonder: to keep relations visible, to keep tensions at the forefront, and to inhabit thresholds where questions about distinctions can be entertained without being shut off because they do not answer clearly to the crisis at hand.

FROM CRISIS TO WONDER

Attending to such futures and to the work of creating their preconditions is a difficult task when we confront concrete images of the effects of the global water crisis: barren landscapes with cracked soil, children drinking from muddy ponds, women walking kilometers with water containers on their heads. Those images circulate through television, the internet, and print and are usually accompanied by pronouncements about the magnitude of the crisis. In 2016, for example, the World Economic Forum polled a group of 750 "decision-makers and experts" from the business world to ask them about the most impactful challenges facing humanity.[11] The respondents listed the global water crisis as the number-one global threat, followed by failure to mitigate and adapt to the effects of climate change and the threat of weapons of mass destruction (World Economic Forum 2016).[12] This same sense of crisis was on the minds of the activists protesting at the forum in Mexico City in 2006. Once their bottles were silenced, representatives from Brazil, South Africa, the United States, and Bolivia waited for the security guards to disperse and then addressed those of us who remained in the lobby. One after the other, the speakers told their audience about water's finite nature. They spoke about the radically asymmetric ways in which that finitude is experienced depending on people's geographic location, ethnicity, class, and gender. They explained the dramatic effects of increases and/or decreases in water flows on species loss, salinization, desertification, erosion, and the drowning or dehydration of multiple forms of life. Without exception, all the speakers ended their speeches with one prescription: the only way out of the global water crisis was recognizing the human right to water and rejecting its commodification.

Notions of crisis, like the one described by groups as different as grassroots protestors and participants at the World Economic Forum, carry with them a particular philosophy of time. They ignite desires to know the genesis of a crisis and hopes to find its timely resolution through historical pivot points. It is not surprising, then, that the task of defining the turning points when things went wrong and, by extension, the moments of transformation when, in theory, things can go back to how they should have been (Roitman 2013: 10–12) elicits all sorts of struggles over the legitimacy and adequacy of "solutions" and those who propose them. While there are multiple angles from which one could analyze the adequacy of those solutions, I am interested in something different from adequacy (Maurer 2005).

Once there is some diagnosis of a solution to a crisis, like the idea of recognizing water as a human right, what happens? We find one answer to this question among my collaborators. As it turns out, once they return to their offices from international meetings and technical workshops, the future again seems uncontrollable and any ultimate solution to the water crisis that seemed workable now appears inadequate. For instance, despite having been framed as an opposition, human rights and commodities go back to looking increasingly alike. And yet the precariousness of the human rights "solution" does not annihilate my collaborator's intentions, nor does it put them in a state of agonistic cynicism. What they do is find ways to retool not only their knowledge but their expectations (Riles 2013). At a time when the magnitude of the water crisis could override any sense of purpose, they find in their technical devices the openings they hope for; human rights acquire new forms and their relation to commodities becomes a knot waiting to be undone. This complex dynamic in a time of water crisis posed an important methodological question for my project: From what kind of ethnographic positioning should I study these devices and people's relations to them? And how do I conceptualize these devices as ethnographic objects?

If some ethnographic moments result in the ethnographers' dazzle (Strathern 1999: 10–11), these devices unleashed something different for me. Marilyn Strathern describes the dazzle as resulting from a particular ethnographic encounter that remained with her for a long time. The mesmerizing sense that encounter unleashed was due to the urge to interpret an unfamiliar observation; a lack of familiarity ignited a lasting search for elucidation. But, as Strathern notes, in anthropology we do not experience the same sense of dazzle with practices or forms of knowledge that are familiar to us because we presume to already know what they are about. During my fieldwork I was not caught by an unfamiliar object. To the contrary, the devices that people brought to my attention were fairly familiar figures, the kinds of objects that we hear about in newscasts and from activists opposing capitalist forms of exploitation. Thus, my focus on those devices did not emerge from an urge to elucidate the unfamiliar. Instead, it grew from another kind of disposition, something more akin to being unsure and hesitant about their place in the world. I came to the project having heard about these devices as world-closing artifacts, but my interlocutors saw them as possibility-creating tools. This conflict made me hesitate. So rather than ignore that hesitation, I turned it into an analytic and affective modality from which to analyze my ethnographic material. I gloss that hesitation as wonder, and use it as a resource to open up contradictory ethnographic objects for joyful exploration. I decided that if technocracy is commonly imagined as a "wonder-killer" I would purposefully engage it as a potentially wonder-inducing ethnographic object.

I want to suggest that wonder, that condition where it becomes imperative to think carefully about things that were presumed totally ordinary, and for that reason self-evident (Rubenstein 2006: 12), is a more generative disposition than crisis to analyze how people like my interlocutors perform political work from the future anterior. It is important to remember that this sense of wonder is not a positive disposition of awe and acceptance. It is closer to curiosity and puzzlement and can bleed into dismay. I am referring to the sense of wonder that we experience when we find ourselves pondering something, unsure of its ultimate significance, ambivalent about its actual implications, willing to take an unexpected direction but concerned about the possible implications of doing so. In this sense, wonder opens up the familiarity of what seems straightforward.

Used in this sense, wonder works both as noun and verb (Swaab 2012). It is passion and thing. It signals an object that amazes and a transitive response that leaves one unsettled. Objects of wonder have "a questioning and questing aspect" (Hepburn 1980: 27). They demand a certain duration so that doubt and confusion can endure long enough to allow qualitative leaps and contradictions in our sense-making. When presented with a conundrum, rather than renouncing or ignoring it, wonder allows an expansion of time, making it possible to dwell in what seems unreasonable—such as a list challenging the physical borders of water.

The devices I study here had that effect on me. They created doubt and concern in my imagination as they claimed to turn water into a human right via a mathematical formula that instills equilibrium and harmony in society (chapter 1, "Formula"); to make the right to water affordable by effacing the subject and celebrating consumption practices of statistically abstracted households (chapter 2, "Index"); to undo the separation between subject and object by attending to the liquidity of water (chapter 3, "List"); and to create a political community by gathering promises rather than incorporating subjects (chapter 4, "Pact"). Used in these ways, the devices I study place liberal ideals about individuals and nature at the border between the acceptable and the unacceptable. They make nature and human dignity mundane, as they translate virtues and values into the dry normality of technocracy. They make the sublime measurable, the sacred regulatable. For that reason, they may seem sacrilegious, doing more than they should, translating things that are supposed to be untranslatable. And yet, at the same time, they ignite passions, trust, and maybe even some hope. When put together in a group, these devices resemble a collection of oddities, a set of objects that challenges our familiar assumptions without being formally authorized to do so. Collectively, these devices do lively metaphysical work while subsumed under the bureaucratic morass of the technical. It is from this position of wonder that I invite you to engage with the devices in this book.

FOUR TECHNOLEGAL DEVICES

Starting from an epistemic mood of wonder, each of the following chapters explores a particular device, gesturing toward its diffuse future, to its engagement with the obstacles of the present, and to the ways in which it activates traces of the past.

Chapter 1, "Formula," examines the work of economic regulators as they calculate the price of water for human consumption. It zooms into the ways in which mathematical calculations become the acts whereby the ethics of human rights are elucidated. Regulators ponder their legal and humanitarian commitments when they navigate the numeric demands of pricing water in a way that excludes profits. This chapter shows how the morality of the profit/rights opposition is translated into a metaphysics of harmony and equilibrium. For regulators, if the variables in a formula are balanced, society will also be. This continuity, suspicious and magical at once, grants regulators space to affirm the ideal of universal access to water from within their technical calculations.

Chapter 2, "Index," shows the unexpected connections between changing consumption patterns in Costa Rican households and the cyclical adjustment of water prices to enact the World Health Organization's prescription that if water is to be an affordable human right, households should pay no more than 3 percent of their monthly income for it. Despite directing their humanistic efforts to making water accessible to the poor, the price adjustments that regulators calculate depend on an economic indicator, the consumer price index (CPI), that targets changing consumption practices across society. Thus, they have slowly reoriented the reach of the human right to water to the things that occupy the home, in the process statistically dissipating the specificities of the human bodies they originally wanted to protect. The result is that the mathematical world of human rights is inhabited by beets, pantyhose, and other commodities, rather than by subjects affirming the intrinsic dignity of their personhood. It is as if the future of the human itself dissipated into humanitarian air.

While in the first two chapters the difference between a human right and a commodity is economically elucidated, the next chapter investigates questions of legal definitions. In chapter 3, "List," I analyze how that definitional challenge takes legislators to the material borders of water, to its very substance. Focusing on the opposition of Costa Rica's Libertarian Party to the recognition of the human right to water, this chapter shows how for Libertarians the materiality of water sets the limits of the (im)possible. Through their procedural maneuvers, Libertarians have composed a wondrous list of water bodies that, they argue, would be covered by linking the recognition of water as a human right to its classification as a public good. Proponents of the reform have argued that the two are indivisible, that the human right to water implies its fundamental recognition as a public good. In their incursion into different forms of materialisms, the Libertarians come to challenge the very possibility of using categories such as public and private to domesticate the morphological indocility of H_2O. To the activists pushing for legal reforms, such material wonderings are ridiculous—nothing but irrational tactics that cannot be taken seriously. Yet, by staying close to that list, we see surprising affinities between "materialisms" of the new wave and Libertarian tactical ontologies. Through that convergence new physical worlds are being implicitly invoked.

Chapter 4, "Pact," shifts to Brazil to examine the all-encompassing character of an initiative called the Water Pact. Here I expand the question of legal obligation to explore efforts made by the Assembléia Legislativa do Ceará (Legislative Assembly of Ceará) in northeastern Brazil to create affective commitments beyond the law. In the Water Pact a group of activists, government officials, and consultants enlist people's capacity to care for water to create an aggregate that, according to its promoters, would have the capacity to transform society's sense of shared responsibilities over water and ensure its universal access. This pact is a form of political aggregation that differs from classic liberal forms, such as Leviathan, which are organized under the premise of belonging. The Water Pact gathers thousands of participants, but does not demand their membership. It is a mechanism to aggregate public promises, and it is predicated on the capacity of a promise to bind people together. With the pact, the organizers hope to expand the meaning and forms that collectives can take—suggesting, on the one hand, a downsizing of the subject to the promise she makes, and on the other, an upsizing of the types of political collectives promises can generate.

Together, these devices can be imagined as a juxtaposition of preconditions to futures that are not calculable from the present. They test new logics and retest old ones in order to remain open to the uncertainty of what the future may carry. Each device constitutes something of a collective attempt, an awkward juncture in which temporalities, utopian imaginations, and pragmatic tactics implode to craft what my collaborators imagine as "vigilant everyday practice": a commitment to the politics of one's expertise. Together, these devices invite a renewed understanding of things we take for granted—a reexamination of our existing worlds and their political categories, through eyes open to wonder. Such an analytic embraces the seemingly monstrous, the mundane, and the surprising in our existing politico-economic repertoires. Perhaps at a time when we are confronting a crisis of our own liberal dreams, reclaiming the wonder in ordinary technolegal procedures can be a generative practice.

1 FORMULA

After I arrived in Cocles, a small town on Costa Rica's Caribbean coast, I spent my first three days shadowing Alvaro. At the time, Alvaro had worked for five years with the Asociación Administradora de Acueducto (community association for aqueduct management, ASADA), which is responsible for providing water to the nearly 150 households in and around Cocles. Community aqueduct organizations like this exist all throughout Costa Rica. They have been promoted by the state since the 1950s as a way to involve local residents in the management of development projects such as aqueducts, roads, and schools. The future of the Cocles ASADA, however, is uncertain. In 2016 the utility that is their legal umbrella confirmed they were going forward with plans to build a large-scale water line to connect Cocles and all the community aqueducts in this region into a single infrastructure. But until that happens, the ASADA continues to be responsible for supplying water to Cocles residents. Alvaro is the ASADA's *fontanero.* The word fontanero comes from fountain, the original source of water in Roman cities. Today the word names those professionals responsible for installing and maintaining the pipes, valves, and engines that move water from wells to mainlines to people's houses. On the first day I shadowed him, Alvaro wore jeans, rubber boots, a yellow T-shirt, and a white cloth hat, of the kind used by farmers in Costa Rica. He carried a black bag on his left shoulder. His right arm was free to grab his *cutachilla*, or "little knife," as he affectionately called the machete hanging from his hip. Alvaro uses his cutachilla to clear the plants that grow on top of the water meters he is responsible for reading.

Our circuit begins by checking a water meter about a kilometer from where we meet. After getting off our bikes, Alvaro kneels, and using a metal tool that looks like an old bottle opener, he pulls up a lid from the ground. Once the lid is out of the way, his hand goes into the hole and pulls up another lid, this time a smaller one. I lean over and see a mechanical meter tracking the flow of water. Its small numbers are rotating, just as they do in an odometer, increasing in magnitude until they reach zero and start over again. "*Están consumiendo agua,*" Alvaro explains. They are consuming water. I immediately imagine women cooking, children brushing their teeth, teenagers washing their parent's motorcycle, grandparents watering plants, sisters washing clothes, nieces preparing *el fresco* (fruit drink). I know that my gendered speculations are most likely to be accurate.

"Each time you open the lid you are in for a surprise," Alvaro says. As he said this, I anticipated he was going to tell me that sometimes the numbers move very quickly or that meters often get stuck. But Alvaro talks about something else. "You never know what you will find in the little caves where the meters sit. Sometimes they are covered with mud just like this one." Alvaro glides his finger slowly over the transparent plastic protecting the mechanism that constitutes the meter. He has done this thousands of times; his body knows exactly how much pressure to exert and how slowly to slide his finger. He continues, "Other times they have turned into ant nests. But not all ants are the same. You can find a nest of *hormigas locas* [crazy ants], the ones that move a lot, like crazy, without a clear direction, but those don't bite you. The ones you need to be careful with are the long, red ants. Those jump to get you, just like these ones!" With his machete, Alvaro disrupts what is left of the nest on the bottom of the meter's little cave.

Alvaro carries a magnifying glass in his bag. His eyesight has become limited following a couple of bad accidents. The worst was when he was setting up a chlorination pump and a hose exploded, splashing chlorine all over his face and eyes. He was blind for a month. Now he can see again, but not very well. The magnifying glass helps him see the numbers he has to read. He could use the pair of reading glasses that *la Caja*, as Costa Rica's public health system is popularly called, gave him while he was in treatment. But Alvaro sweats a lot doing his job and this, combined with the heat and humidity of the Caribbean coast, ends up fogging his lenses. The magnifying glass is much more practical.

I offer to help and Alvaro lets me read the water meter as he prepares to write the numbers I will dictate. After feeling with his hand the contents of his bag, he grabs what used to be a transparent plastic bag that is now of a milky color and pulls from it a thick stack of index cards. Each card corresponds to a particular meter; he looks for the one belonging to the house

Figure 1.1. Sign at entry to the Cocles ASADA.

we are visiting. Leaning over the ground, I read to him the numbers on the meter and he writes them on the last row of a column of numbers that expands monthly, one row at a time. Once I stand up he shows me how they use the numbers we have collected. They subtract the old number, the one on the previous line, from the new one he has just written. The difference between the two is the *consumo de agua* (water consumption). After his explanation, we continue with the rounds and check sixteen more meters. By 10:30 AM the sun is too high and the temperature too hot; Alvaro recommends we ride back to the ASADA.

When we arrive, the door of the ASADA is open, but no one is inside. We call Alex's name and he comes out from his home, next door. We talk a little, and then begin the procedure to turn the numbers we have collected into bills. Alex, a college student who also works as manager of the ASADA, opens an Excel file with a long list of names next to a column with a series of "user numbers." Alvaro passes the information he wrote on the index cards along to Alex, who enters it on the spreadsheet. Once all numbers are tabulated, Alex sends the file to an accounting firm that has supported the Cocles ASADA since they joined a regional federation of community aqueducts.

The firm takes the information people like Alvaro and Alex collect and multiplies it by the price of a cubic meter of water, a figure that is set by ARESEP (Autoridad Reguladora de los Servicios Públicos), Costa Rica's public services regulation authority. The accounting firm Alvaro's community aqueduct has hired then sends the results of their calculations back to Alex and also introduces them into a national electronic payment system that allows people to pay their bills at any bank, grocery store, or pharmacy. The sweaty act of monitoring water consumption, pushing ants away, moving index cards in and out of a bag, wiping the sweat off your magnifying glass, and dictating numbers travels back and forth through Excel spreadsheets, internet routes, accounting systems, and paper documents until it becomes a bill, the price people pay for having their human right to water delivered. This procedure takes place once a month, every month. The numbers that Alvaro and I collected are interlocked with the pricing formula regulators in ARESEP use to set the prices of water. Alvaro determines how many liters per month have been consumed. ARESEP determines how much each of those liters, at any second, cost. Laws, histories, and institutions connect those worlds; the monthly calculation of prices reenacts that connection cyclically. Worlds separated by 220 kilometers are entwined in largely invisible ways, through the calculations performed by an Excel sheet.

Because of their day-to-day experience at work, Alvaro and Alex are aware of the power ARESEP has, and for that reason they are also curious as to how that agency works and how they make their numbers. Even though ARESEP is a powerful force that shapes innumerable daily economic transactions—they set the prices of gas, public transportation, electricity, and water—the agency remains mostly unnoticed when everyday people think about water. To the contrary, ARESEP regulators constantly think about the relations between the "everyday people" whose rights as users of public services they must protect, the water providers that are in direct contact with those users, and the prices they produce. This is especially important in the case of water, which is, in the words of ARESEP personnel, the most fundamental public service the state can supply.

This chapter takes you to ARESEP's offices to explore the formula behind the key ingredient in the alchemy that turns Alex and Alvaro's numbers into a final number on a bill. I was captured by how, despite their apparent distance from places like Cocles, the decisions ARESEP makes have such direct power over the life of water among people in Cocles and elsewhere in Costa Rica. It is often in places like ARESEP, distant from particular pipes, wells, and chlorination procedures, that the life of water is crucially shaped. Alvaro and Alex are very aware of this since any deviation from what ARESEP determines is the legal price of water can get them in deep legal trouble, through either accusations of privatization or allegations of water corruption.

To trace these distant, yet intimate, connections I analyze the pricing formula regulators use in ARESEP to create a bifurcation between a bad price and a good price, a price that reproduces commodification and a price that affirms human rights. Such a distinction, as I learned, is produced with people like Alex, Alvaro, and the families they service in mind, even if no one in ARESEP knows them personally. Thinking of the Alvaros and Alexes of Costa Rica, regulators work to create a price that excludes any notion of profits. They do so to distinguish between a humanitarian treatment of water and a commodified one. From the point of view of regulators, and largely but not only for legal reasons, if a utility is generating profits out of its service delivery, it can be said to have commodified water. If it is not generating profits, knowingly or not, the utility is following a more humanitarian approach. What is interesting is how this distinction is made at the same time that people pay for water according to the cost of its purification, distribution, and management. Thus, though my interlocutors see themselves as drawing a distinction, at the end of the day, once that distinction is set, they realize that it affirms water as both a human right and a commodity, and creates the need for another way to make their desired contrast evident.

To show how this process unfolds, this chapter follows the work regulators do to shape the algebraic relations between water, citizens, humanitarian injunctions, and economic ideologies of profit. I say algebraic relations because these are relations mathematically expressed in a formula. ARESEP's pricing formula is a mathematical proposition that embodies metaphysical assumptions of balance, harmony, and equilibrium as enshrined in the law that regulators are charged with implementing. As I will show, for regulators the relations between the variables in the pricing formula have direct effects in the world. If the relations between variables are harmonious and equilibrated, regulators see that very same balance and equilibrium in the relations between citizens and utilities, and ultimately in society as a whole. All of this material-semiotic potential requires that we attend to the formula slowly, thinking carefully about what we might intuitively consider as technically ordinary.[1] As a device that makes differences matter, we need to attend to this formula in its thick moral histories and world-making capabilities. To do so, I will first take you to ARESEP, its public hearings, and its political place in Costa Rica. Next, I trace the legal principles that inform the formula ARESEP regulators work with to show how, inspired by ideas of harmony and equilibrium, regulators calibrate their formula and its variables to enact larger social imaginaries that go beyond water. I then trace a controversy over a recent attempt to change the variable that deals with profit-making and show how the quasi-event of this shift threatens historical ways of allocating financial and humanitarian responsibility among water providers. I call this a quasi-event because its occurrence is not fully realized in the sense that the shift never officially happens. As we see in other chapters in this book, a quasi-event unleashes peculiar effects; in this case it generates an eventful and quotidian technical struggle about how profits determine the humanitarian nature of water.

TWO SOFIAS, ONE FORMULA

Rather than discuss "the economy" as a coherent entity unto itself, most Costa Ricans primarily talk about prices, routinely commenting on how expensive things are and how high *el costo de la vida* (the cost of life) is. Comparing prices against their available income, against each other, and against what they are willing to pay, people are often frustrated about their limited resources. But beyond the immediacy of everyday consumption, prices are also collective objects of concern. Newspapers, politicians, and activists refer to them as entities that affect social relations. Through their intimate and public lives, prices draw attention to fundamental questions about the nature and role of the state, the meaning of the notion of an economic community, and the limits of financial tools for quantifying the value of substances as fundamental to life as water.

As is the case elsewhere, the prices people in Costa Rica encounter in their monthly bills are nonnegotiable (see figure 1.2). The formalization of

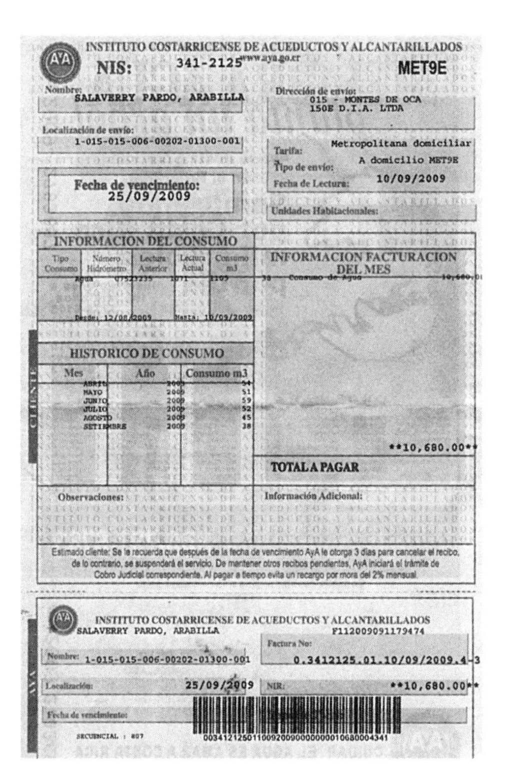

INSTITUTO COSTARRICENSE DE ACUEDUCTOS Y ALCANTARILLADOS
NIS: 341-2125 www.aya.go.cr MET9E

Nombre: SALAVERRY PARDO, ARABILLA
Dirección de envío: 015 - MONTES DE OCA
150E D.I.A. LTDA

Localización de envío: 1-015-015-006-00202-01300-001
Tarifa: Metropolitana domiciliar
Tipo de envío: A domicilio MET9E
Fecha de Lectura: 10/09/2009

Fecha de vencimiento: 25/09/2009
Unidades Habitacionales:

INFORMACION DEL CONSUMO

Tipo Consumo	Número Hidrómetro	Lectura Anterior	Lectura Actual	Consumo m3
Agua	07523235	1071	1109	38

Desde: 12/08/2009 Hasta: 10/09/2009

INFORMACION FACTURACION DEL MES
Consumo de Agua 10,680.00

HISTORICO DE CONSUMO

Mes	Año	Consumo m3
ABRIL	2009	54
MAYO	2009	51
JUNIO	2009	59
JULIO	2009	52
AGOSTO	2009	45
SETIEMBRE	2009	38

10,680.00
TOTAL A PAGAR

CLIENTE

Observaciones:
Información Adicional:

Estimado cliente: Se le recuerda que después de la fecha de vencimiento AyA le otorga 3 días para cancelar el recibo, de lo contrario, se suspenderá el servicio. De mantener otros recibos pendientes, AyA iniciará el trámite de Cobro Judicial correspondiente. Al pagar a tiempo evita un recargo por mora del 2% mensual.

INSTITUTO COSTARRICENSE DE ACUEDUCTOS Y ALCANTARILLADOS
SALAVERRY PARDO, ARABILLA F112009091179474
Nombre: 1-015-015-006-00202-01300-001
Factura No: 0.3412125.01.10/09/2009.4-3
Localización: 25/09/2009
NIR: **10,680.00**
Fecha de vencimiento:
SECUENCIAL : 807
00341212501100920090000000010680004341

AYA

Figure 1.2. Water services paper bill from AyA.

large parts of the economy has reduced the wiggle room people have when they visit pharmacies (many of which are chains headquartered in other Latin American countries), buy from large grocery stores (most of which are now subsidiaries of Walmart), or pay for public services, all of which are regulated by ARESEP.

Regulators at ARESEP cherish the prices they produce, in large part because they see them as having important capabilities that people lack.[2] Prices have the ability to rank, order, and connect across scales and domains of social life (Guyer 2004). It would be impossible to imagine a person who could simultaneously affect how cash-flow projections are made in a water company, influence the decision about whether or not to replace a broken water meter in a poor neighborhood, foster conversations within families on how to maximize their monthly budget, allow credit card companies to add automatic payments to people's accounts, and an infinite quantity of other daily practices and strategizing exercises. The prominence prices have in everyday life conceals the fact that, despite their solid image as a cohesive entity, they are constituted by myriad elements that tend to remain out of sight (Guyer 2009: 205). The patterns by which those elements are brought together and put in relation with each other tell us a lot about what is a political community, how the state intervenes in it, and what is a shared resource. As compositional entities, prices allow people to communicate the unsaid and they open spaces for unexpected reinvention.

I began my journey to learn the elements that constitute the prices that Alex and Alvaro pass on to people in Cocles by physically going to ARESEP. In 2008 I attended my first *audiencia pública* (public hearing) there. After meandering around the streets of San José to avoid traffic and make it to the 5:00 PM hearing, I finally arrived at the agency's headquarters, located in the western part of the city. A security guard showed me the entryway to an auditorium that had been added to the building many years after its initial construction during one of the many remodels it had gone through. From its creation in the 1990s until shortly after 2010, ARESEP occupied that same apartment building. Over that period, its administrators performed all sorts of architectural modifications, not only to make space for the growing number of employees but also to match changing ideas of what a public office should look like.[3]

In the auditorium, the hearing was about to start. With a capacity of about one hundred people, that afternoon there were no more than forty attendees between ARESEP employees, public servants, and utility personnel. Public meetings like this are a central piece of modernist state-making. They allow public officials to assemble symbols of authority and citizens to perform their assigned roles (Li 2007). They are also strategic and carefully crafted displays of knowledge and ignorance about technical issues (Mathews 2008). But these meetings can also turn into tournaments of political skill where people challenge their expected roles as they convene for information, consultation, or organization purposes (Alexander 2017). The lawmakers who created ARESEP imagined these meetings as a means to increase transparency and bring citizens closer to the state. This particular audiencia pública was held to collect public feedback on the latest petition by the largest water utility in the country, AyA, to increase the price they charge for water services by an alarming 40 percent.[4]

From the front stage, a young man in a business suit formally guided the audience through a legal and administrative ritual whose high point was a presentation by Sofia, a member of what at the time was the Water and Environment Department (WED) of ARESEP.[5] Sofia is trained in business administration and has worked at ARESEP for close to two decades. When I met her, she was in her mid-thirties and was one of the few women holding technical positions in her department. At the meeting, Sofia was in charge of the PowerPoint presentation because she had been appointed to lead the evaluation of the latest price increase request received by the agency. Her presentation had been prerecorded, and while she sat in the audience, we saw her enlarged image onscreen giving a fifteen-minute introduction to price regulation. Sofia informed attendees about how the agency analyzes the legal and technical propriety of the petitions that water utilities, all of which are state or municipal entities, regularly submit. Most of her talk, as one would expect, revolved around the regulatory methodologies her department follows. But in between analyses of demand elasticity, depreciation rates, and efficiency, Sofia insisted on the responsibility "all" people have to guarantee the implementation of the human right to water. This strong reference surprised me.

My surprise came in part from the fact that it was only in 2010, two years after I met Sofia, that the United Nations General Assembly passed Resolution 64/292 recognizing the right to safe and clean water for drinking and sanitation as an essential condition for the full enjoyment of life and all other human rights. What at the moment I had temporarily forgotten was that this recognition was the culmination of more than three decades of international discussions. And yet, while welcome, the official recognition of water as a human right by the UN General Assembly did not significantly alter the thinking of most regulators. They, and most Costa Ricans for that matter, already recognized the existence of a universal human right to water as something of a natural fact, a self-evident truth. One of the first formal definitnions of the right to water was adopted by the UN in 2003, well before the General Assembly resolution. That definition already used a term that was especially meaningful to Sofia and her colleagues. That definition noted that, among other things, if water is to be a human right it needs to be *affordable*.

After her prerecorded presentation was over, Sofia walked to the stage and projected a slide with the regulatory formula WED uses to calculate water prices (see equation 1.1). ARESEP adopted this formula in the 1990s following a recommendation by a consultant from the Pan-American Health Organization hired to modernize the methodologies the agency used. Explaining how the variables in the formula made their humanitarian responsibilities concrete, Sofia informed her audience about the possibilities

$$X = O + A + D + R$$

Equation 1.1. Formula used by Costa Rica's public service regulation authority to set the price of water.

she saw in the elements that make up the prices they produce. A human right to water was not something determined in courts, it was something that resided in their calculations. Their formula, as any equation, was in reality a story about the relations between different entities. And as any story does, this one had material effects in the world.

Sofia's braiding of the price formula with the affordability of human rights was neither romantic idealism nor superficial talk. As I later learned, she wholeheartedly embraces human rights as a powerful instrument for directing public attention to political and economic inequalities.[6] Yet, by drawing attention to how human rights infuse the relations between variables she was also revealing that in these relations, as in any relation, there was tension—in this case, between two logics of political and technical intervention: financial capitalism and humanitarian ethics. On the one hand, moral sentiments and ethical concerns for the Other are intrinsic to the very idiom of human rights. These concerns are put at the center of governing regimes, particularly of those regimes designed to help the poor or the disadvantaged (Fassin 2012: 1). On the other hand, there is the logic of finance, which includes "all aspects of the management of money, or other assets . . . as a means of raising capital" (Maurer 2012: 185). This logic extends well beyond the limits of financial markets into all sorts of economic relations organized around credit, debt, and revenue.

And yet, the tension between these two logics is not necessarily a mismatch between two incompatible philosophies of value. For Sofia, it was more like a puzzle that challenged her and her colleagues to arrange their numbers appropriately, calculate them ethically, and organize them harmoniously. These are two different flows of ideas and preoccupations that coexist inseparably. Sofia's job consists of giving the tension arising from their coexistence the right intensity so that their prices can stand an ethicality test, an assessment of how they will affect the lives of others.

These tensions between different logics of technical intervention shape the relations between the variables in the pricing formula regulators use.

The variables allow for intertwining those logics but also for creating separations between them. Thus, as I will show, the humanitarian implications of financial theories do not have to be analyzed externally only, as when we go to the "real world" to document the effects the usage of those formulas have. We can also trace some of the implications of those logics inside the formula, through the relations between variables, just as regulators do. Such evaluations constitute a processual point of reference against which regulators redefine the limits of appropriate (technical) action (Faubion 2010). The worksheets, mathematical models, and legal resolutions people use when they calculate their variables are, on the one hand, means to reveal the humanitarian standing of water and, on the other, instruments to sharpen their own ethical awareness of the decisions they make (Keane 2010: 72). Worksheets, models, and legal resolutions connect the intimacy of one's ethical evaluation of everyday work to larger political and economic contexts.

COSTA RICA AS AN OBJECT OF ECONOMIC AND REGULATORY HISTORY

During its golden welfarist history (1950–80), Costa Rica's activist state strongly participated in economic production matters and elevated the living standards of most of its population through universal social policies (Martínez Franzoni and Sánchez-Ancochea 2013). The abolition of the army in the late 1940s, the nationalization of the banking system, and a constitutional reform making schooling through the ninth year mandatory, accompanied by strong labor protections and a payroll tax that funded the social security and public health systems, made Costa Rica, before the 1980s oil crisis, the most universal and least stratified welfare regime in Latin America.[7]

After its welfare heyday, starting in the 1980s, Costa Rica was caught by the global "neoliberal" wave. Understood as a preference for market mechanisms to deal with collective issues, a push for opening the economy to foreign investment and a liberalization of currencies, the neoliberal mantle wrapped the well-grounded core of social institutions that Costa Rica's population continues to depend on (Vargas Solis 2011). Part neoliberal, part welfarist, Costa Rica's technocratic cadres and political elites advanced a hybrid agenda that, despite not following a radical program, introduced enough reforms to widen the gap between the richest and poorest citizens. By 2012, the formerly least stratified welfare regime of Latin America found its richest citizens with an income 14.5 times larger than its poorest citizens (CEPAL 2013).

Despite the neoliberal hype of the 1990s, Costa Rica did not ultimately transfer its strategic public utilities—electricity, water, or telecommunications—to private control, regardless of the many attempts by different administrations to do so. A series of citizens' mobilizations, union demonstrations, and actions by some politicians who remained committed to the country's "exceptionalism" as a welfare state contained the privatizing hype that triumphed elsewhere in the region. Nevertheless, to keep with the economic fashion of the 1990s, and in characteristic Costa Rican fashion, the country's Legislative Assembly created ARESEP as an autonomous regulatory agency charged with regulating the state itself through its public utilities, as opposed to regulating private providers, as was the case in the rest of Latin America.[8] This power quickly turned ARESEP into a key player in the structuring of Costa Rica's common resources and public sector.[9]

ARESEP was created by law in 1996, at the height of the era of structural adjustment and regulatory capitalism. The agency matured along with the privatizing trend that swept Latin America during the 2000s, when the state was reimagined as a regulatory entity responsible for setting clear "rules of the game" for private players and for promoting "markets" as preferred mechanisms for allocating resources and resolving social struggles. In that context, agencies like ARESEP were assigned the responsibility of mediating between corporations, citizens, and the state via "technical" decisions.

It would be misguided, however, to assume that because all strategic utilities continue to be public in Costa Rica, the prices ARESEP sets for them automatically abide by principles of common economic welfare. Utilities have undergone a process of financialization due to the particular theories, accounting standards, and mathematical models that have become normative knowledge among economically trained personnel around the world (Thrift 2005). Hence, the public or private nature of a utility has lost traction as an index of distinct legal and economic logics. To understand the ideas of society and the values that undergird a utility, it is necessary to examine in detail its financial and administrative practices; its status as a public or private entity is no longer enough.[10]

Following global pressures and due to their own creativity, regulators in ARESEP have experimented with a variety of tweaks to their pricing methodologies. Some of these experiments have been based on the material and infrastructural transformations of the services being paid for. In water provision, an important shift occurred when utilities moved from charging fixed rates to a system of charges proportional to the quantity of water consumed following the installation of individual meters outside people's homes (for insightful studies of the everyday experiences of accessing and paying for water, see Anand 2017; von Schnitzler 2016). Other shifts in pricing logics do not correspond to infrastructural changes but to legal and economic shifts. For instance, in the 1990s there was a global push toward full cost recovery, the idea that all costs utilities incur should be paid for, preferably by the consumer. This was for all practical purposes an attempt to turn away from subsidies. More recently, there has been a new shift toward humanitarianism and universal rights, whereby the attention goes into issues of universal access and affordability. But beyond their specific political implications, these historical shifts show the capacity of a pricing device, a formula that has remained constant, to braid a broad range of assumptions about society, the state, and justice, even as context shifts.

TECHNOLEGAL METAPHYSICS: HARMONY AND EQUILIBRIUM

In their daily work, regulators in ARESEP believe that, once formally adopted, the prices they calculate will disseminate through society the values imbued in them throughout the process of their creation. The first description of this relation between prices and society I heard was given to me by Don Marcos, a former director of WED, temporarily reassigned to lead the "future projects" team while the agency was "reengineered" in 2009. Don Marcos is a long-standing public servant. Sofia describes him as the living history of regulation in Costa Rica. He started working on water issues in the 1970s when ARESEP was not even in the imagination of lawmakers and regulation was done from within utilities themselves. Don Marcos is affable. He embodies what Costa Ricans imagine as a good bureaucrat: a person who gives you confidence and knows what he is talking about but never makes you feel that he knows much more than you do. Don Marcos's speech is measured; he never seems to rush. When he discussed with me the impact his work has on Costa Rica's population, he referred to a continuity between the two by saying, "We have in our hands the most social of all public services; that is why any change in our methods will be a change in society, mostly for the poorest users." For him, the price on a bill is a performative encounter between citizens and utilities.[11] A bill is a way of summarizing people's relations with society at a given time (Hart 2007).[12] People's bills convey some of the values that organize the political and economic communities they live in. Regulators see prices as indices of sociality and reject any reductive definition that presents them as simplistic reflections of intrinsic value (cf. Kopytoff 1986). As Sofia put it, prices never capture the real value of water, but they approximate as closely as possible the economic dimension of the social relations that guarantee its access.

A few weeks after Don Marcos shared his views on prices and society, Sofia and I had one of our first "interview-like" conversations. She came to the conference room where I was waiting for her with her arms full of gifts for me. She was carrying brochures, children's books, a card game on water conservation, a calendar, and a few booklets explaining the mission of ARESEP. All the materials had been produced by the User Relations Department. That department regularly produces these kinds of materials so that regulators can distribute them during public meetings or other outreach events they organize. The documents are effective pedagogical artifacts that explain things like the agency's mission and instruct citizens on how important it is to save water and pay your monthly bills on time. Among the documents Sofia handed me was a short booklet with blue covers, a copy of the Ley de la Autoridad Reguladora de los Servicios Públicos Número 7593 (Law of the Public Service Regulation Authority Number 7593). Passed in 1996, this law established ARESEP's mission and to this day continues to be a source that regulators cite routinely, most of them from memory.

Days later I sat again with Sofia, this time in her cubicle. We had scheduled a follow-up conversation in which she was going to help me understand how their work at ARESEP unfolds. As we began she picked up a booklet like the one she had given me. Hers was overflowing with yellow Post-its marking the heavily used pages. She started flipping through the pages with the confidence afforded by having done so many times. Once she found the page she was searching for she looked up to confirm that I had also located it in my own copy. She turned her booklet toward me, pointed to the middle of the page, and said, "These are the reasons why we exist, this is what we have to do."

Sofia was physically and conceptually pointing to the fundamental principles she and her colleagues use when dealing with their pricing formula. She proceeded to read Article 4 of the law, which mandates that ARESEP is responsible for "harmonizing the interests of consumers, users, and providers of public services and for seeking equilibrium between the needs of users and the interests of providers." She mentioned this without much emphasis, with the familiarity of something that has become unremarkable. But for me the obligation to create harmony and seek equilibrium was an enchanting task full of religious undertones and utopian desires. Her words brought together an assortment of accounting theories, economic belief systems, and metaphysical assumptions about relationality.

In addition to the theological undertones, this fantastic injunction to create a world in harmony and equilibrium has two implications. On the one hand, it specifies the kinds of social relations that ARESEP is responsible for fostering—the ends of its work. On the other hand, it also establishes harmony and equilibrium as properties that the agency's mathematical methods should also exhibit. Harmony and equilibrium are means to their ends. With this dual character, as means and as ends, harmony and equilibrium constitute a technolegal metaphysics of sociality that touches all calculative and pricing activity in ARESEP.

As we continued discussing the numeric expression of the call to effect harmony and equilibrium in the world, Sofia scribbled in my notebook an explanation of the pricing formula she had presented at the public hearing where I first met her (see figure 1.3). This time, Sofia disaggregated the equation to explain how, in the search for harmony and equilibrium, the financial income and the expenses of a utility were her bottom-line concerns. The variables in the formula allowed her to trace different spheres of action in the utility: operations, investments, future plans, and so on. But most of these are usually fairly straightforward, even if they require verification. Where things get tricky is with the variable that captures that difference between income and expenses, noted as R in the original equation.

As I supplemented her notes with my own, Sofia led me through stories about the adjustments and tweaks regulators do to each variable in their "real-life" dealings with utilities. In contrast to her presentation of the formula during the public hearing, this time she told me about all sorts of contradictions and proposed changes. For each variable, there were long series of reformulations, waves of consultants recommending changes to computation methods, styles of exercising authority by new directors ap-

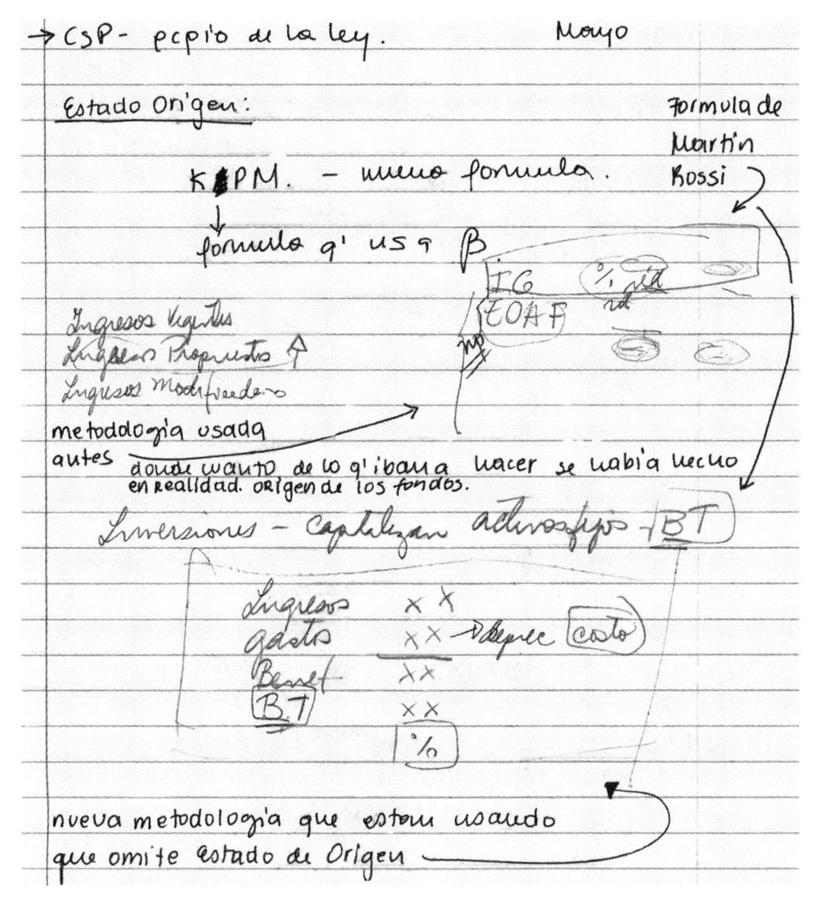

Figure 1.3. Sofia's new rendering of the pricing formula that emphasizes the question of profits.

pointed to her department, and quibbles among coworkers about the role regulators played in society. The formula was a lively space. Despite its apparent continuity, since it was adopted in the 1990s, it was the space where a lot of changing ideas came to be discussed.

As she continued, Sofia explained how she really worries about the problem of profits. Sofia knows that for her fellow citizens profit-making marks the difference between a human right and a commodity. A commodity is something you can profit from; a human right is not. In Costa Rica, all utilities providing water services are prohibited by law from profiting from their activities. This prohibition comes from the legal principle of *servicio al costo* (not-for-profit-service), also included in Law 7593. This principle states that water prices should be designed only to recover costs and can never be used to generate profits. Sofia and her colleagues assume that without their policing of the application of the principle of servicio al costo, utilities would seek opportunities to accumulate surpluses, something that is not just legally prohibited but is viewed by most in the agency as unethical. But there is a twist. The principle of servicio al costo does allow utilities to generate a "competitive" surplus to raise adequate resources to improve the quality of their services. Thus, servicio al costo precludes profits but not surplus, transforming the fundamental question of human rights into a very technical difference: the distinction between surplus and profits in the calculation of a variable in a formula.

To elucidate the difference between surplus and profit, Sofia spends a lot of energy wrestling with the financial difference between income and expense, and analyzing how that difference shapes the magnitude of R, the variable that stands for surplus in their pricing formula. That variable is technically called "development yield" to differentiate it from what for-profit operations call return on investment or ROI. These different names are necessary because the very same formula is used in other parts of the world to regulate private utilities.

Thus, when utilities in Costa Rica request ARESEP's authorization to collect more income via increased prices, regulators carefully assess whether the request will create any surplus. If a surplus is produced, regulators have to apply the principle of servicio al costo to judge whether it is an acceptable R (development yield) or a form of disguised profits, an unacceptable R (ROI or profit). To know if R is one or the other, development yield or profit, it needs to be seen in the context of the other variables. That conditional dependence precludes the possibility of defining R in absolute terms, that is, by a fixed magnitude. Consider a hypothetical example: If they assessed the absolute magnitude of R, regulators could decide that a utility with a surplus of, say, 1,000,000 *colones*[13] is unethical and another utility with a surplus of 100,000 is ethical. In that case, the mere magnitude of the surplus, in and of itself, would guide the decision. But what if in the first case, 1,000,000 was a difference of 3 percent between income and expenses, and 100,000 a difference of 15 percent? We can see that the absolute magnitude of the number is misleading. The proportion between income and expenses—that is, the relation between R and the other variables—is how regulators determine R's ethical character.

This lack of an absolute magnitude to assess R's ethicality makes it unwieldy and pliable. It creates the need for detailed policing of its magnitude on a case-by-case basis. For that reason, regulators always search for a hidden surplus as a potential subterfuge for profits. The right magnitude of R—displaying the right proportionality—is a moving target for which a vigilant eye and a continuous implementation of the principles of harmony and equilibrium are necessary. These contests around the magnitude of R as the right proportionality are one of the very concrete numeric locations where regulators cyclically re-create the preconditions of the future histories of water. In those numeric locations they have to ponder the ability of utilities to collect enough financial resources to do their job and fulfill their obligation to protect users from being charged excessively. What today they classify as an appropriate R will shape what seems reasonable for utilities and users, or not, in the future when new requests to increase the price of water come in for evaluation. Via discussions about R's magnitude, regulators create the preconditions that render harmony and equilibrium present. These conversations also determine both the present and future humanitarian (im)morality of a utility's operations and their own place in securing said harmony and equilibrium in the social body.

LEDGERS, THEOLOGIANS, AND MARKETS

When scholars trace the origins of harmony and equilibrium as metaphysical assumptions, they often turn to theological and legal principles (Agamben 2011). Theologically, harmony and equilibrium derive from heavenly order and godly omnipotence. Legally, harmony and equilibrium inspire the doctrines of checks and balances and rule of law behind liberal legal systems. But in the particular regulatory work unfolding in ARESEP, harmony and equilibrium cannot be traced directly to God or exclusively to legal principles. Instead, they are mediated by the numeric relations that go into producing a price.

In Latin America's legal history, harmony and equilibrium have figured as paradigms of a legal ideology that has sometimes dismissed conflict and the language of rights and preferred the more benevolent language of agreements, trade-offs, and compromise (Nader 1990). As Nader noted a while ago, the quest for harmony is a search for balance that domesticates conflict and often reproduces the status quo. Recent innovations such as legal restitution, alternative conflict resolution, and truth commissions are also geared to restore a moral balance interrupted by illegalities and horrors of different sorts. It is tempting to focus on legal practices and postcolonial encounters to elucidate how harmony figures in the struggle for the human right to water in Costa Rica. I will sidestep that route, however, in order to trace another history of harmony and equilibrium, one that is less known but that shapes more directly the formula regulators use. That is the history of the relation between harmony and equilibrium, the ledger, and the market. This history is directly relevant because in 2009 the ledger and the market were at the center of a political controversy over the precise numeric form of the variable R.

One of the most widespread histories of the ledger, the double-entry bookkeeping technology, puts its invention in the Middle Ages, and explains it as a key moment when the preoccupation with theological harmony acquired numeric form. At first sight, the ledger was a document that recorded income and expenses and the differences between them. It was an accounting tool to track credits and debits, resolving them in a number that expressed the relation between them, whether they were balanced or one exceeded the other. But its meaning went further. The ledger was also taken as evidence of the virtue of economic exchanges. Theologians and religious officials rejected avarice and the excessive accumulation of wealth as challenges to godly authority and used the ledger to verify whether a merchant followed those principles or not. If a merchant accumulated large amounts of wealth, as shown in the final balance of the ledger, that person would be guilty of avarice and greed. If, on the other hand, the final balance between credits and debits in a merchant's ledger was zero, that number "conjured up both the scales of justice and the symmetry of God's world" (Poovey 1998: 54–55). This arithmetic offset of a number by its opposite, of credits by debits, was the embodiment of harmony and equilibrium proper. It was evidence of balance in God's creation.

At a larger scale, this godly symmetry was also expected of global mercantilist trade. Imbalances between imports and exports weakened the godly order colonial powers claimed justified their imperial excursions into Asia, Africa, and the Americas (Finkelstein 2000). Whether it was at the level of the individual merchant or at the level of global trade, the importance of a final numeric balance did not reside in its referential meaning, nor in its formal precision. What mattered was that said balance, the proportional offset of credits by debits, preserved the symmetry and order of God's creation.

Amid the concerns over harmony between virtuous merchants and global trade, there was another unit of economic activity that generated existential debates around bookkeeping, equilibrium, and the future of Europe: the price. Medieval thinkers, such as Thomas Aquinas, elaborated the notion of a "just price." For him, the adequacy of a just price was not determined by how accurately that number captured the components—costs—behind the production of a particular good. Instead, prices were taken as a form of recompense (Hamouda and Price 1997) to be granted to those who aligned with God's order in earth. Thus, as a measure, prices were everyday tools to harmonize faith and social reality.

Later on, the Salamanca School of Economics in Spain, a group of theologians concerned with the virtue and fairness of economic action since the fifteenth century, argued that a just price could also emerge from market interactions (Melé 1999: 175). Their contribution was singular because they built their discussions of justice and morality on the separation of price-setting from the representatives of divine authority on earth. Their argument was that "natural" prices emerge out of the valuation that actors do of a certain good and that, as an extension of divine order, those prices are necessarily fair. They noted how pairing the value of a good to the cost of its production, something only knowable by the producer, opened space for the exploitation of poorer segments of the population. The necessary virtuous proportionality between cost of production and price, between producers and consumers, could easily be corrupted. To prevent such moral decay, the Salamanca School recommended disentangling kings, priests, and producers from price-setting and letting the "impersonal" forces of the market—which were, like everything else, ultimately godly forces—do the work of setting them (Elegido 2009). Classical economists such as Adam Smith, David Ricardo, and John Stuart Mill secularized and combined these views later on. They understood the composition of prices as a summation of the costs of production of a good. Once publicized in the marketplace those prices would take on a dynamic of their own (Cetina 2006: 554).

The ideas of the Salamanca School are unexpectedly relevant to my discussion of the formula regulators use to set the price of water.[14] For one, the resurgence of market-centric economics since the 1970s revived interest in the Salamanca School's works by enthusiasts of political economic thought from the Austrian and Hayekian schools. Libertarian economists often invoke the Salamanca School's belief in the market as the cradle of justice. The theologians were concerned with the monetarian impact of the influx of gold and silver from the colonies to Spain and with the status of

Amerindians, their bodies, labor, and property, in the new imperial landscape. They argued for the recognition of people in the Americas as godly subjects and for free market relations, and not kingship, as the best way to organize property and wealth. Their thought is used today by believers in free markets and deregulation to explain the *longue durée* of the connection between the market, freedom, and justice. Their revival is so strong that economic tourism agencies organize tours for market advocates to go to Salamanca and visit the chambers where theologians wrote about the market. And yet something that is lost in those celebratory tours and in the popularized version of market economics is that the school's enchantment with the market was also an attempt to better understand its limits.

Amid their analyses of the virtues of the market, the Salamanca theologians identified important exceptions. In particular, many theologians argued that goods that were considered "vital" or of "prime necessity" deserved a "legal price" fixed by public authorities (Melé 1999: 183). Thus, something else that connects the Salamanca School with Sofia's interest in income and expenses six centuries later is a shared interest in goods that need to be priced by a public authority because of their fundamental role in sustaining life. These are goods that, in Sofia's words, "need to be kept outside of the volatility of the market." The foundations set by merchants, theologians, and moral and political philosophers through their theorizations of ledgers, prices, profits, and markets figure directly, albeit implicitly, in contemporary discussions around how equilibrium, harmony, and human rights shape the relations between R and the other variables in the pricing formula ARESEP uses.

PRICES WITHOUT PROFITS

"Prices are signals," Martín categorically told me one morning when we were chatting. Martín is, without a doubt, the most polemical economist on the Water and Environment (WED) team at ARESEP. That morning he politely pulled a chair next to his desk so that I could sit in his cubicle to conduct our interview. As we conversed, he sipped water from a disposable bottle. He told me he had purchased that bottle before coming to work that morning. Martín was taking better care of his health, so he was drinking water more regularly during the day. As we continued talking, he turned his bottle into a pedagogical resource. "How much do you think this liter of water costs? Do you think it is cheap or expensive?" he asked me. I thought

I could foresee his line of reasoning. I expected him to continue by referring to Adam Smith's famous "diamond-water paradox." Smith was famously perplexed by the fact that although humans cannot exist without water and we can easily exist without diamonds, when compared unit per unit, the value of diamonds was vastly greater than that of water. The paradox is one of those popularized pieces of Smith's thought that travel widely across expert circles.

Before I could respond, Martín quickly answered his own question by going in a different direction from what I had expected. "Too much!" he cried with the intensity that characterizes him. "This liter of water costs too much, it is as expensive as a liter of gasoline! And if you walk around this office you see people all have their bottle with them. Why, then, should public utilities not charge more for the water they supply? People are buying bottled water, they can pay for it. Why should the state subsidize people that can pay for the luxury of bottled water?" I was somewhat stumped by Martín's ideas. He was generally right. At the time, a liter of gasoline was about US$1.93 and a liter of bottled water US$1.90. The water that utilities provide is generally of good quality. There is no public health reason to avoid it except in some community aqueducts where water sources have been polluted or they are having problems with their chlorination process. And yet a large segment of the urban population purchases bottled water regularly (for the politics of bottled water consumption see Pacheco-Vega 2015; Wilk 2006).

Martín is inspired by libertarian and neoliberal ideas, and when he shares his thoughts he often incites strong reactions. He is prone to create controversies and, as he readily admits, enjoys doing so. He once told me with a thick grin, "because of my beliefs, I am not the most popular person here." I could see the basis for his reputation when, after explaining why he was utterly convinced that water should be managed for profit and through markets, he gave me two documents to study. One was from the Cato Institute, the libertarian think tank based in Washington, D.C., and the other was from the World Bank. Both documents argued that subsidies were causing the world's water crisis and that privatized markets were the only solution for securing the future of the infrastructures and utilities managed.

Martín continued instructing me on his views on economy, society, and particularly prices. He believes there is no better communicative invention than prices. Paraphrasing Hayekian thought on the problem of information in planned economies, Martín subscribes to a naturalized view in mainstream economic circles: good prices come from markets.[15] From that context, they are able to perform their communicative magic. In Martín's and Hayek's view, the magic of prices is that they communicate information to large numbers of people in ways that centralized planning tools, such as regulations, never can. But this association between prices and markets that Martín is so fond of is far from being generally accepted. In the case of water, many regulators, along with most citizens, are suspicious of what market pricing can accomplish.

I should note, however, that the general wariness of market prices among ARESEP employees and citizens is not a rejection of prices in general. In Costa Rica, people readily accept that they should pay for water. What they are suspicious of are prices produced by markets, because people generally understand markets as spaces that conceal intentions to extract excessive profits. Here, *mercantilización*, a word I commonly heard from activists and water professionals, is relevant. As a term, mercantilización indexes markets, commodities, and profits all at once. It refers to the exchange of a certain good through market transactions designed to extract excessive, and thus questionable, profits. Mercantilización goes beyond reciprocal exchanges of value to signal an intention to extract wealth and exploit others with little consideration of their well-being.

For regulators, however, mercantilización is a technically obscure concept. It has no specific meaning they can engage numerically, and thus they tend to not think about it in their daily calculative routines. Politically, the story is different. The meaning of mercantilización is clear. When activists and community organizations mobilize against the commodification of water and argue for its genuine treatment as a human right, they are in fact arguing against its mercantilización. They argue that the prices charged for water cannot follow market rationalities, take advantage of people, and generate profits for a utility. The notion of mercantilización captures all of this. It is so politically charged that regulators fear the possibility of the media reporting that their decisions are pushing water in that direction.

Martín, however, is not scared of mercantilización. As a follower of his namesake, Martin Hayek, Martín believes the market is equipped to correct for its own excesses. Yet, regardless of his heartfelt Hayekian proclivities, by working at ARESEP Martín has no choice but to produce prices that are far from the free market semiotic wonders he admires. His prices are regulated entities, "legal prices" as Salamanca scholars called them, subject to strict regulations and administrative oversight by the bureaucracy. The

$$X = O + A + D + R$$

where

X = Cost of service delivery
O = Operation costs
A = Administration costs
D = Devaluation
R = Development yield (or return on investment)

Equation 1.2. Formal mathematical expression of the formula, and its variables, that the Costa Rican public service regulation authority uses to set the price of water.

irony was never lost on me. It is not surprising that Martín's ideas do not go over smoothly among his colleagues, who for the most part steer away from Hayekian ideologies.

Months later, I saw Martín representing the Water Department at another sparsely attended public hearing, this time held in Heredia's local chapter of the Chamber of Commerce.[16] The chamber is located in a residential house in the center of town. It has also been remodeled multiple times. The hearing was held in an atrium that at times served as a covered parking lot. That afternoon, rows of white plastic chairs had been put in place for the audience and a large projection screen set in the front next to a table with white and blue tablecloths. The same young man from ARESEP, with what seemed to me the same business suit, was again officiating at the ritual. Martín gave the main technical presentation, equivalent to the one Sofia had given previously. Martín was in charge of leading the team evaluating the price increase request that Heredia's municipal utility had submitted a few weeks earlier. Just like Sofia, he began his presentation with WED's pricing formula and a careful definition of each variable (see equation 1.2). He then explained how the relations between variables had to reflect harmony and equilibrium if the relations between utilities and users were to also embody those qualities. As he moved along, he cited different articles of ARESEP's law. Unlike Sofia, however, he never mentioned the human right to water. Through his disaggregation of the formula, the audience learned about the financial logic by which operation costs, administrative costs, and depreciation rates are kept in equilibrium with a justifiable surplus (development yield), represented by R.

The development yield (R), that tricky mathematical difference that Sofia first presented at the public hearing and then scribbled on my field notebook, again occupied us. After Martín's explanation, the financial manager of the utility requesting the price increase for which the hearing had been organized took the stage. She showed a series of slides with the financial projections the utility had submitted for ARESEP's evaluation. All of her tables had a red balance showing how, at current prices, the utility would soon fall into serious deficit. From the utility's perspective, those red figures embodied the imbalance that made it reasonable for ARESEP to increase the development yield (R), thereby authorizing an increase in the price of water for Heredia's neighborhoods.

This imbalance, a red number that reflected a negative difference between expenses and income, is an important symbol. Without further articulation it invokes an unjust situation, a challenge to (godly) harmony. Income and expenses are unbalanced. Following the logic of the ledger, this is the justification utilities generally need to convince ARESEP of the urgency of the price increase requests they are petitioning for.

Regulators cannot take that red balance for granted, though. They are obliged to investigate its accuracy by looking at the accounting practices behind it. In Martín's words that afternoon, regulators have the obligation to verify the accuracy of these figures by scrutinizing their connection to the "real world." Throughout the efforts involved in that investigation of the financial statements that utilities attach to their petitions to show their connection to the real world, regulators can adjust the intensity of their oversight. They can be more rigid or lax depending on how accurate they deem the connection between the statements and the real to be. As a result of that "reality" check, regulators can adjust the variable R up or down as long as they keep it between 3 and 7 percent. This variation grants utilities more or less income, as regulators see fit. To this day, utilities criticize ARESEP for being too rigid in their analyses and too bureaucratic in their procedures and for never understanding their real needs. Sometimes utilities even accuse regulators of being stuck in time, in the era of the welfare state, rather than adapting to the epoch of economic efficiency.

After public hearings like this, and once they are back at their cubicles reviewing the utilities' accounting practices, regulators begin evaluating the proportionality of R—to determine whether its magnitude reflects the ethical assessment they make of what a utility deserves. As we saw, R does not have an absolute magnitude that reflects harmony and permanently distinguishes between profit and surplus. To be ethical, its magnitude has to embody an acceptable proportion between income and expenses. Thus, there is always the risk of opening the door for mercantilización if R is not calculated correctly. But if regulators are too stringent and do not give utilities enough financial wiggle room by adopting a very narrow R, they run the risk of financially strangling them. So being overzealous can cause problems for water users as well. The ambiguity of R, as something that can create surplus without calling it profits, turns it into something of a nominalist trick, a variable whose determination is a game of names that requires constant assessment to establish the propriety of its magnitude in relation to the other variables in the formula: operation costs (O), administrative costs (A), and devaluation (D). R can tilt toward being an indicator of the virtuous efficiency of utilities or a covert form of profits that needs to be disciplined. Setting the numeric propriety of R is thus the very juncture where the human right to water, understood as something that cannot be profited from, takes numeric form. Regulators all over the world constantly face situations like this, junctures where a seemingly technical shift has the potential to push things over a certain boundary and change them radically, even if they mostly stay the same. These are moments when a difference needs to be effected, even if it is not absolute and everlasting. This is how effecting differences sets the preconditions of the future.

PERFORMING SOCIETY

Don Marcos, the former director of WED we met earlier, was the first to refer to the continuity existing between the work regulators do in their offices and the world that exists outside of those offices. Recall how Don Marcos told me that they had in their hands "the most social of all public services" and that any change in their formula was in fact a change in society, particularly in the lives of the poor. It is not difficult to see why he said that. WED sets the price all Costa Ricans pay for water. Remember how Alex and Alvaro's work in Cocles is ultimately shaped by ARESEP prices. The decisions regulators make literally touch everybody's pockets. Allowing profit-making from water service provision would be a large-scale transformation. But Don Marcos's observation that a change in their formula was in fact a change in society had a deeper meaning. It was not only that regulators are aware of the consequences of their actions; he was conveying something about the intimacy between their formulas and the world they live in. This became clear to me days later when Sofia, also in pedagogical mode, shared more materials with me.

After getting some coffee and finding space in a meeting room, Sofia and I were talking about the connection between their methods and society—the real world, as Martín calls it. I recalled to her Don Marcos's observation that any change in their regulatory methodologies is a change in the lives of people, particularly the poor, because they are affected more significantly by any change in water prices. At one point in our conversation Sofia thought it was best for me to do some reading before continuing our discussion and she proceeded to send me an electronic copy of a manual she had used for an e-learning course she had taken in 2008. The course was organized by the Asociación de Entes Reguladores de Agua Potable y Saneamiento de las Américas (Latin American Association of Water and Sanitation Regulatory Entities, or ADERASA). People in ARESEP often mentioned this association when comparing their accomplishments to regulatory innovations in other countries in Latin America. The association was created in 2001, with Costa Rica as one of its eight founding members. The group was tasked with organizing training events, convening political fora to identify regional positions, and providing technical expertise to regulatory authorities throughout the continent. The course Sofia took was part of a broader distance-learning program led by economists that started in 2006.

That night, in my apartment, I opened the document and was stumped by what I read. The first page of the document laid the foundations of the practice of economic regulation. In its first section, the manual began by stating that "the costs of a regulated company depend on the type of regulation established" (Asociación de Entes Reguladores de Agua Potable y Saneamiento de las Américas 2007: 3). As I read this section, it was impossible not to smile. Pages and pages of sophisticated social studies of finance condensed into a matter-of-fact foundational statement for any regulator: economic facts are performed by the knowledge through which we produce them (Callon 2007; Mitchell 2005). The statement bluntly expressed that the costs of any utility will always, and only, be those costs regulators choose, through their methodologies, to count as such.

Later, when I commented on how striking that sentence was, Sofia underlined that even something like a cost is never an external fact preexisting their calculations. Like so many other things in her job, a cost is brought into existence by the principles that agencies like ARESEP use to shape their formulas and the relations between their variables. Sofia remained a little amused by my fascination, and I don't think I was ever able to explain clearly to her why I found this so interesting. In schooling me on the performativity of economic calculation, Sofia wanted me to understand the power regulators have in their hands. They reshape the world, even if they seldom step outside of their cubicles to do so.

This understanding of their work is what fills the gap one might see between regulatory circles and the lifeworlds that unfold elsewhere, in places like Cocles. This understanding of the performativity of their methods explains why for regulators creating harmony and equilibrium in a formula, as demanded by the law, can be seen as an act of creating a society in harmony and equilibrium. In previous historical moments the connection between world and formula was provided by godly omnipotence and the precepts of biblical scripture. In the present, the connection was effected by the performative, and taken for granted, power of accounting, economics, and the principles of economic regulation. Recognizing that one has this power is sobering. So it is not surprising that Sofia is cautious about any new directions that power could take them. Methodological changes are always risky terrain.

FROM ACCOUNTING TO ECONOMICS

The tasks involved in overseeing the work of utilities, and particularly the magnitude of R, open spaces to enact different regulatory temperaments. These are forms of calibrating how intense and detailed or lax and general their evaluations are. The regulatory approach ARESEP had followed for years required them to examine R in detail. Sofia explained the affective charge of this approach by saying that they had historically behaved like *peseteros*. This expression derives from the term *peseta*, which refers to the Costa Rican twenty-five-cent coin, removed from circulation in the 1980s because of its loss of value due to inflation. A person is called "pesetera" when she or he obsessively chases and wants to account for every single penny and how it has been used. In English, an equivalent term is "penny pincher." The pesetera attitude is something Sofia confesses is a bit excessive, but also necessary when one is responsible for equilibrium and harmony in society. The ethics of public service require one to think in those terms, and even more so in the case of water because it is a human right, she told me during one of our conversations. The pesetero temperament results from the historical particularities of the Costa Rican state as much as from the grip that economic ideologies hold on regulators' everyday calculations. The tension between those ideologies and the state's welfarist legacy became most visible when ARESEP began considering a potential change from an *accounting approach* to an *economic approach* to regulation. The struggle around this change revealed how the long histories of the relation of prices to ledgers and markets that I mentioned before acquired renewed saliency.

In 2009, the new director of the Water Department in ARESEP, Alfonso, proposed moving away from the old accounting approach and shifting to an economic approach to regulation. Despite being new to the agency, Alfonso was assigned to Don Marcos's position soon after he was hired. He arrived as part of the team of close collaborators the new head of ARESEP brought when he was appointed by Costa Rica's president. The idea of moving from accounting to economics was part of a modernizing initiative launched by the new government. One of the consequences of that initiative was the relocation of ARESEP's building. After heated controversy in the media, the agency moved from the old remodeled apartment building it owned to an expensive building they rented in the newest area of the city. Their new building had shiny white floors, a glass façade, and required one to pay a toll to get to it by taking the only highway in the country that has been privatized.

Rumors about the impending methodological change that Alfonso was promoting compounded an already crisp political environment. In 2007, via a controversial referendum, the country ratified a free trade agreement with the United States and the rest of the Central American countries. The agreement divided the country and for many was a major hit to what was left of the welfarist state apparatus. Opponents were convinced the trade agreement would lead to the privatization of all utilities, flexibilization of the work force, erasure of labor protections, and bankruptcy of many agricultural industries. One of the commitments the country made in the agreement was to open the telecommunications sector, which by law was a state monopoly, to private investment. For Costa Rica's telecommunications industry to be attractive to private corporations, Congress recategorized it from public service to regular commercial service. This classificatory shift exempted companies providing the service from the principle of servicio al costo that I explained earlier (see the Technolegal Metaphysics section).[17] This structural change implied that from 2007 onward private, for-profit corporations could offer cell phone, fixed phone, and internet services, all of which had previously been provided by a single state-owned utility. ARESEP remained in charge of regulating the newly born industry, though a semi-independent regulatory office was created. But, at the time, and from the perspective of Sofia's team, it became impossible to ignore the telecom team's efforts to figure the appropriate mechanisms to regulate private and public entities simultaneously. A shift from an accounting to an economic approach to regulation would have similar effects in the regulation of water, even if it did not involve a full declassification of water as a public service. It would be a mathematical shift that would carry consequences for how society dealt with profits in water provision.

The practical consequences of the change were clear. The shift from an accounting to an economic approach would tether regulators' daily calculations of R to one of the basic assumptions behind the idea of markets: that they are aggregate entities. As a popular icon of the modern economic imaginary, even if an inaccurate one, the invisible hand of the market is taken as an aggregate of moral and economic forces counterbalancing each other (Smith 1966). According to this view, those forces engage in a rhythmic push-and-pull dynamic until they reach a form of equilibrium. In this view of the market there is an issue of quantity at stake. One consumer and one producer trading with each other do not constitute a market. Only the summation of multiple consumers and producers can be analytically abstracted into the geographic space, extended network of relations, or Cartesian graph that we use to envision what markets are. What all of these images of the market share, however, is the vision of equilibrium as a result of the aggregate forces of supply and demand.

The approach that dominated ARESEP's methodological discussions up to that point followed the accounting logic of double-entry bookkeeping to track the magnitude of the balance and potential profits (R) in a single utility's operations, not in the aggregate. Because it focuses on financial statements, people refer to this as an accounting approach. As we saw, when it was invented, the ledger offered insight into the virtue of a merchant's practices and had the capacity to combine legal, economic, and theological traditions into numeric forms to appraise deference to godly harmony. In the ledger, the significance of the final numeric balance between credit and debit does not depend on the precise volume of income and expenses. Its ethical significance depends on whether credits and debits offset each other. If the final balance is zero, then the merchant's activities have been virtuous. This arithmetic equilibrium between credit and debit is specific to an individual merchant, or in this case to an individual utility's R.

Something else that made the accounting approach particularly valuable for regulators was that, beyond assessing financial records, it created opportunities to investigate the practices behind inscriptions of income and expenses (the ledger), the reality Martín mentioned. Regulators took it as their responsibility to talk to a utility's personnel about their reports, look for cues of dangerously creative accounting, and make their regulatory acts an interpersonal affair. Phone conversations, email exchanges, and face-to-face meetings allowed them to gauge the virtue of a utility and its deference to its financial/humanitarian obligations.

After decades following this accounting approach, meticulously policing the balance between a utility's income and expenses and setting an R that did not allow for profits, the Water Department began to consider breaking with this established approach. Alfonso, the new director, wanted to move the agency toward aggregate analyses, toward thinking of all utilities as something of a "market" rather than continuing to focus on the numbers of individual utilities. The practical implication of the shift was a reorientation of the attention previously given to accounting reports and balance sheets. In the new approach they would determine a fixed "industry-wide" R that would apply to all utilities regardless of what their ledgers showed. In this new approach, an adequate R (development yield) would be standardized. Instead of regulators having the discretion to increase or decrease R according to a utility's doings, its magnitude would be automatically and universally set. This example shows how the process of financialization of water unfolds, even as water remains a public good.

In line with their pesetero attitude and with the accounting approach to regulation it represented, regulators historically controlled R and kept it somewhere between 3 and 7 percent. That variability allowed regulators to move up or down in order to keep the relations between variables equilibrated and the principle of servicio al costo alive, as expressed by this formula (see equation 1.3). In the formula, the size of R is proportional, mathematically and aesthetically. It contributes to balance and equilibrium by being adjusted to the magnitude of other variables.

If ARESEP moved to an economic approach to regulation, R would have

$$X = O + A + D + R \text{ (flexible between 3 and 7 percent)}$$

Equation 1.3. A harmonious and equilibrated water-pricing formula.

$$X = O + A + D + R \text{ (fixed at 5 percent)}$$

Equation 1.4. The pricing formula in disharmony and lacking equilibrium.

a fixed value standardized for all utilities, municipalities, and ASADAs. At the time, the magnitude that was discussed was 5 percent. In that scenario, R would be independent of any other variables in the formula, as shown (see equation 1.4). This change would render the equilibrium-infusing effects of the pesetero attitude and the accounting approach moot. R would stand on its own, unconnected, uninterested, and in defiance of what the magnitude of the other variables signaled. That could easily bring about a slippage of development yield into profits, as an automatic 5 percent R would be granted even when utilities had a surplus.

Until 2009, a flexible R (between 3 and 7 percent) had been a crucial improvisational space where regulators choreographed their intimate knowledge of a utility's ledger. A flexible R had been critical for their capacity to imbue their formula with ethical qualities—harmony, equilibrium, and non-mercantilización. R allowed them to mold, according to that image, the relations among water users, utilities, and society, all from their cubicles. If the proposed change occurred, the contours of those ethical qualities would be radically transformed. The nonharmonious and unequilibrated formula would also manifest in a society in disequilibrium.

With this possible change to R came a wealth of rumors about major ideological changes in regulatory calculation grammars. Sofia was especially concerned that the discretion to control utilities they had enjoyed in the past would cease to exist. The space they had in the past to evaluate the connection of financial accounts to the real world would shrink. Many regulators connected the rumors to their fear of losing control over the public nature of water. Stories about veiled mercantilización circulated, and activist groups, community aqueduct organizations, and users started voicing their apprehension in meetings, workshops, and public hearings. Furthermore, publications in the media and discussions on social networking sites connected these possible changes with the demands of the free trade agreement with the United States and Central America ratified in 2007. Opponents argued that one of the objectives of that agreement was to privatize water and open the water export business to supply countries in the Global North.

But the emotional and political turbulence inside and outside WED did not diminish Alfonso's enthusiasm for transitioning from an accounting approach, following the aesthetic of the balanced ledger, to an economic approach guided by the idealized logic of market aggregation. A new future seemed to be on the horizon. One member of Sofia's team conceptualized the shift as an attempt to move from accounting as evidence of prudence to aggregate measures as indications of efficiency. R would become a symbol of financial freedom for utilities. Finally, Costa Rica would catch up with countries that people like Don Marcos saw as much more advanced in regulatory issues, such as Chile and Argentina.[18] Other regulators spoke of the change as a "neoliberalizing" measure that would introduce disguised profits and fracture their long-standing commitment to servicio al costo and affordable water access consistent with human rights obligations. The economic approach would decouple yield (R) from balance sheets, closing off possibilities for steering, correcting, and rewarding utilities in their dynamic search for harmony and equilibrium and creating a standardized financial rent. The formula would no longer be producing a bifurcation between two prices on the basis of whether they generated profits or not. Their previous tactic to preclude profiting at all costs would no longer be the point of separation between a price that aligned with a commodity and a price that aligned with a human right. Or at least that tactic could no longer focus on R. Their formula, while staying the same, would be different. As a device it would braid different histories and as a result shape the world in different ways.

While the shift from an accounting to an economic approach was not to the liking of many regulators, those who enthusiastically supported it justified its merits in terms of standardized responsibility. Alfonso thought the shift would transfer the obligation to self-police financial sustainability to the utilities, freeing regulators to focus on issues of service quality, another fundamental characteristic of the human right to water. This responsibility rationale was based on a financial habit they had detected in utilities' accountings. Through the years, water companies had grown accustomed to operating very close to, or sometimes in, deficit as a way to justify their petitions to augment their development yield (R). Recall how, during the hearing that Martín led, the financial manager of the utility requesting the price increase showed a series of financial projections painting a dire future in which the company would fall into deficit. She was not the only one using that tactic. Almost all of the files that petition for a price increase have similar figures. This perpetual proximity to deficit also was, as Sofia's e-learning manual noted, the performative effect of how regulators used their formula to account for costs and police profits. It was the world-making effect of the capacity of this device to entice certain responses. Transcending the representational value of numbers and interlacing calculative pasts and futures had historically resulted in a choreography of equilibrium, harmony, and not-for-profit pricing that valued unbalanced financial statements and in some ways made deficits desirable. The promoters of the change argued that a standardized R would finally break this pattern and force utilities to adopt a more entrepreneurial attitude—to be "less paternalistic," as one of them said.

Sofia was not pleased with the implications of the change. She frequently explained that within her team she was known as a champion of users, especially the poorest segments of the population. Arguments of fairness, for her, were significant as long as they considered how "the poorest of the poor" were affected by technical decisions. She saw in the new approach a disguised pull away from regulators' substantive commitment to affordability and low prices. Their inability to adjust the development yield in each specific case would, Sofia contended, inevitably increase the price charged to users, something they had been very careful about, since high prices were one of the biggest threats to securing universal access to the human right to water.

Martín, unsurprisingly, welcomed the new approach. He thought it was about time water utilities grew out of their habit of being policed and started being more responsible for their own actions. For him, utilities needed to become "financially smarter," catch up with their obligation to manage themselves more efficiently, and stop relying on *Papá Estado* (Father State) to guide their decisions. "It is a good idea to move toward a fixed development yield and let financial balance be something that companies worry about, not us," he said. Rather than state-making projects, for Martín, utilities had to be rethought as entrepreneurial entities resembling subjects whose ingenuity, creativity, and market initiative have to be encouraged. His equality argument followed the rationale that, if all utilities are operating in roughly the same conditions, for some utilities to have certain things recognized as costs while others do not is unfair. The logic should be that all utilities are granted the same level of working capital. That move would include fixing R in the aggregate. Not doing so discriminates against some utilities, making some seem more efficient than others even if they are not.

Resembling the 1990s atmosphere of infatuation with neoliberal economics among mainstream economists and politicians throughout Latin America, at the end of the first decade of the 2000s, a profit-friendly mantle was enveloping regulators' work. New ideas were loosening regulators' historical commitment to the prevention, at any cost, of profits. Discussions on the nature of a fixed R deepened the ideological differences between regulators, exemplified by Martín and Sofia, especially when they addressed how households were going to be affected by the price increases, how companies would suffer if they did not have freedom to invest, and what this redefinition of equilibrium and harmony between elements in the formula would mean for human rights.

A CHANGE THAT NEVER WAS

What for a while seemed an imminent change, eliciting all sorts of anxieties and hopes, never came to fruition. The move from accounting to economics, from a flexible R to a fixed R, never happened. The shift was a quasi-event, not completely unleashing its intended effects, as it was never adopted, and yet it impacted the everyday affects, histories, and practices of people in regulatory offices. In April of 2013, four years after the discussion of the shift started, the front page of the largest newspaper in Costa Rica reported that AyA had accumulated a staggering surplus of about $415 million dollars. The head of AyA explained that delays in an infrastructure renewal program were responsible for this surplus. Yet, on the basis of that surplus, AyA's petition to increase water prices was flatly denied. ARESEP was still steering utilities toward humanitarian and financial virtue through their pesetero approach to ensure harmony and equilibrium between variables. The difference between income and expenses in the ledger still commanded the metaphysics of harmony and equilibrium.

It would be a mistake to completely disregard the rumored change in R, and its lack of adoption, as an exceptional and inconsequential occurrence. Unrealized changes occur regularly in ARESEP as people consider new technical possibilities. The density of preoccupations, calculative experiments, and reflexive assessments among regulators during those periods is not unusual. On the contrary, that intensity constitutes the liveliness and unpredictability of calculation as people use their numbers to shape the worlds they want to live in. Ideas of new theories and methods constantly come through the agency. Consultants, distance learning courses, and new authorities help bring them into regulatory offices. The threads that the pricing formula weaves are constantly being tested. People like Sofia are always attentive to how unrealized changes in the past are unexpectedly picked up later.

Despite its apparent stability, the formula is a lively device of the ethical, financial, and legal explorations of profits and of financial humanitarianism. At the end of the day, those explorations take a specific form, they settle on a distinction, they effect a bifurcation between humanitarian and nonhumanitarian prices that is temporarily emplaced until new ideas seep through and regulators realize that new variables and their relations with other variables need to be revised.

In my most recent meeting with the new director of WED, the third director I have met since starting this research, the focus had changed. Now they are not focusing on R, but on the quality of the services utilities provide. This is unleashing a whole new set of speculations, innovations, and differentiations that deserve an analysis similar to the one I have conducted here. Their commitment to making water universally accessible at an affordable rate but also in the right quality is bringing new questions to their pricing formula, to the story it embodies, and to the effects it unleashes in the world.

CONCLUSION

Alvaro and Alex, the people we met in Cocles, know very little about what happens inside ARESEP. When the agency announces changes in the prices they have to charge to their neighbors, they just get an email detailing the increase and instructing them to inform water users. To comply with that obligation, Alex prints and tapes the email to the office's main door. But since people no longer have to come to the office to pay their bill—they can do so in commercial establishments or banks—very few neighbors notice the announcement. As a result, the first time Cocles residents, or most Costa Ricans for that matter, notice that the price of their human right to water has increased is when they make their monthly payment.

This cleavage between what happens in a community aqueduct and ARESEP's decision making is striking. As we saw throughout this chapter, this is not the case for AyA and the larger utilities, which regulators audit much more closely. But in both cases what is interesting to me about this configuration is that intimate fieldwork in the Cocles community aqueduct or in AyA would not reveal much about how exactly those changes come into being. By focusing on regulators we see how ideological changes take technical form through the variables in a formula. Those variables, and the relations between them, embody particular theories about institutions, behaviors, and matter. By changing a variable in a formula, regulators help unleash a series of events that can reshape the flow of water through pipes, its timely chlorination, repairs to leaks, and myriad "materially" tractable happenings. That change is the intensification of historical threads, institutional contexts, and even personal desires. I have argued that by looking at how regulators wrestle with their numeric formulas, we can precisely trace how broader ideological trends sculpt financial structures of public services and, through them, social collectives. Those changes are the ones we then gloss under greater categories such as structural adjustment and austerity. To confront those macro changes, we need to know how they come about.

The pricing formula regulators use is a device that determines the contents of the bills that 1.7 million Costa Rican households pay every month. It numerically determines a price that can be said to stand for their human right to water. But this is far from being solely a Costa Rican issue. The approach to regulation that ARESEP follows, focusing on controlling the return that utilities get from their operations, is the most traditional and widely used regulatory methodology around the world. Most countries in the Americas, including the United States, follow some variation of this methodology. Throughout regulatory commissions, authorities, agencies, and committees, experts discuss what is an appropriate return for a public service. Not in all cases is return used as a proxy for a human right, but in all cases it helps assess the morality of what the public nature of a service means, and how the frontier between private and public constantly shifts.

If the prices in people's bills account for the relations between individuals and society at a given time, it is important to ask how those in charge of designing numeric figures make sense of those relations. And, more importantly, how do the instruments they use shape those prices? My emphasis on the formula regulators use, rather than on citizens' experiences, intends to highlight the labor regulators perform when they mobilize their methodologies. I have shown the often-overlooked openness and instability of those practices. Unlike accounts of law and finance that black-box technical decisions as obscure, complex, and somewhat mechanical procedures, I have shown how technicality is constituted by ethical, cultural, and mathematical sets of relations in a constant state of potential change. Seen in this light, Sofia's and Martín's engagements with ARESEP's formula requires an ethnographic account that, despite its intensely specific explorations, cannot be reduced to the micro-local. Their everyday work demands an ethnography that engages both the indivisibility of the historical legacies of the formula and their future-making potential. That type of ethnography attends to historical predecessors and the effects of the formula as preconditions of the future.

ARESEP's pricing formula entails the everyday preoccupations, rules, and transgressions through which regulators make sense of ideas that at points seem competing injunctions and at other points seem two sides of the same coin. On the one hand, most regulators have deeply felt commitments to a humanitarian obligation to keep water affordable by excluding profits. On the other hand, they understand that they are setting a price following the logic of finance and in order to generate revenue to keep utilities and community aqueducts running. While competing, these injunctions also coexist, sometimes seeming inseparable, and for that reason inviting people to attempt to separate them. And yet ARESEP's pricing formula allows regulators to make that difference emerge by tilting their methodologies to one side or the other. The relations between variables in the formula is the space where their views on society, the fundamental role of water, and the place of the state in contemporary capitalist formations can be visualized. Their formula is their social theory, and a performative one at that.

This chapter has taken us through the vicissitudes of living and working with a pricing formula and through it putting one's social theory to the test. Those tests include moments of mathematical determination when the difference between a flexible R (3–7 percent) and a fixed R (5 percent) is the difference between a price that is guided by harmony and equilibrium in society and a price that is not. That moment helps people make sense of the difference between a price that precludes profits, thereby helping achieve affordability, and a price that does not. As I have shown, that moment of differentiation might seem trivial as we consider the magnitude of the water struggles and transformations people fight for. But in fact, it is an expansive space where theological histories, economic theories, personal relations between regulators, and international free trade agreements come to matter. Thus, while magnifying the bifurcation this formula produces, I have also described the broader historical and political contexts that come to matter at the precise technical moment regulators effect their differentiations.

Mapping the expansive threads that come to matter at that juncture has taken us on a circuitous tour through ant-filled water meters, legal principles, economic theory, public hearings, and bureaucratic cubicles, so that I can argue for the importance of understanding how potential shifts in the humanitarian and financial worlds we live in are assessed. The apparently unconsummated change of R I have described is one example of the many unrealized modifications that regulators consider and try out. I have suggested that, given its unsuccessful transformation, the change of R is not a full event, yet it is not a nonevent either. The intense discussions, rumors, arguments, and hypothetical numerical exercises to understand a possible change in R are all preconditions of the future. They are happenings that could become significant antecedents; we just cannot know yet if they will. Ultimately, the point is that discarding their significance because we cannot see their effects in the present erases from our analyses the subjunctive worlds in which people live most of their lives in spaces like Costa Rica's public agencies, but also elsewhere. This is exactly what I mean by doing a type of ethnography that engages the indivisibility of historical legacies from their future-making potential.

But the process of creating bifurcations has not stopped. New iterations continue. In this chapter I focused on the very definition of the variables involved in pricing. Once those variables are defined, change comes from elsewhere. The price of water does not remain stable, and other technical questions need to be addressed. A crucial one is how to update those prices to match the changing value of money and economic conditions. The next chapter takes us again to Costa Rica, where another precise technical decision that brings in another set of historical and contextual threads needs to be made. This time, the question is how to think of affordability as a temporal problem.

2 INDEX

El costo de la vida, literally the cost of life, is a fundamental worry for people in Costa Rica and elsewhere. El costo de la vida is an economic concern, but that does not mean it is limited to exchange via money, especially because such exchange is always more than just a financial transaction. El costo de la vida entails one's preoccupations about budgets, decisions to prioritize spending, ways to meet obligations to care for others, and the allocation of time and resources for play. How one uses one's money, how it is distributed, and what it accounts for captures all sorts of emotional, political, and social relations and obligations. Where water figures in the distribution of one's income and how that expense relates to other items is another way to understand water relationally.

As I do with the other devices I trace in this book, in this chapter I explore the ways in which people engage the world once they know that it is relational and entangled. How do they create separations between categories that resist them, and how do their devices help them do so? In this chapter I am particularly interested in what the passage of time does to the work of creating such distinctions when setting the price of water. How do regulators attend to changing financial value and how do they think of those changes as they consider collectives, large numbers of people sharing a common infrastructure and a watery environment? As in the previous chapter, I begin with an ethnographic vignette that gives you a taste of how that preoccupation with affordability is expressed by people working on a community aqueduct. I then take you back to ARESEP, and later to the National Statistics Institute, to examine how an index is used to affirm the humanitarian nature of water, as opposed to its commodified character.

In our examinations of this particular index, the consumer price index (CPI), we travel in history to learn some of its technical precedents and then follow its production and usage in Costa Rica. Those travels show how the consumer price index is a device that creates a difference between human rights and commodities in a surprising way, by displacing the subject of rights and giving primacy to the household of goods. Here we see how a society that is squarely organized around a liberal understanding of the human, as a bearer of rights and a citizen of the republic, relies on purchased goods to determine the humanity of its subjects. This emphasis on objects, particularly purchased ones, requires that we expand our imaginaries of subjectivity and open the door to the multiple ways in which material objects are crucial for the definition of humanitarian reason.[1]

In this new imaginary of humanitarian reason, subjects and objects are blurred, not because of religious beliefs or ontological experiments, but because what counts as the human right to water is answered by looking into what objects inhabit people's houses. This operation of blurring object and subject is one that is numerically accomplished and is organized around the mandate of calculating the affordability of water. Thus, another way of thinking about the work regulators do is as the enactment of the numerical and material ontology of rights. The rest of this chapter explores how this unexpected, though not uncommon, configuration of humanity and rights is brought about. But let's begin this exploration at a workshop with representatives of community aqueducts and water-related NGOs, where affordability is understood as the fundamental humanitarian worry.

WATER: A GIFT OF NATURE AND GOD

While waiting to take the floor at the workshop she was attending, Ana had become unsettled by the words just uttered by her colleagues, a group of community aqueduct and local NGO representatives who had gathered to discuss the meaning of transparency in the implementation of the human right to water. Convened in San José, the capital of Costa Rica, the group met at the headquarters of an NGO founded by a former foreign relations minister who now worked to promote peace and sustainable development internationally. The workshop was called to *validar* (validate) the initial results of a survey that workshop attendees had participated in a couple of months earlier. Among other things, the survey evaluated how thoroughly respondents understood the economic implications of recognizing water as a human right. According to the slide on the screen, 83 percent of the organizations "agreed" to one degree or another that water was an economic good. An air of awkwardness permeated the room. The attendees found themselves embracing economic concepts usually espoused by politicians they distrusted. Even though the attendees dealt with costs, debts, and bank interest rates in their daily activities, speaking about economic goods at a meeting where they normally discussed mobilization tactics was unorthodox.

Ana's turn to speak came after a man from Guanacaste, the driest province in Costa Rica, finished his remarks. In response to the slide on the screen and the idea that most understood that "water is an economic good," he had reminded us that water was also a gift from Nature and ultimately God, a sacred substance that belonged to all, a human right. When Ana stood up to address the group, she said passionately, maybe even with a little bit of irritation, "Yes, *el compañero* [our colleague] is right. God and Nature gave us water. It is a human right. I agree, but it is too bad God didn't put it in all of your houses. Somebody has to bring it there and that has a cost." She continued, "There is no question people have to pay for water. Our ASADAs [community aqueduct associations] cannot work otherwise. The issue is whether people can pay, whether they can afford their bills."[2]

Ana's opinions were legitimized by the fact that she works as a volunteer with an ASADA, a type of community organization responsible for managing small-scale aqueducts under the legal supervision of AyA, Costa Rica's largest public water utility. Her ASADA is located in a small town in Costa Rica's central mountain range, about two hours west of San José. But the notion of volunteering is misleading here. Although she is not paid for her work, Ana is available any day, at any time, to deal with the constant issues that emerge in her ASADA. Her firsthand knowledge about how ASADAs work made her anxious about the pressing economic problems these community organizations face. In order to keep their infrastructure running, buy chemical treatments, and pay for fontaneros (plumbers) to maintain pipes in acceptable condition, managers of ASADAs throughout Costa Rica use creative means to keep alive a sociotechnical system many deem financially unviable. A newspaper article published in July 2015 in *La Nación*, the country's largest newspaper, for example, called attention to how ASASDAs supposedly operated *en completo descontrol* (completely uncontrolled) because AyA, the entity legally responsible for their oversight, had little knowledge about their finances. But despite the journalist's attempt to discredit ASADAs as particularly disordered exceptions in an otherwise healthy and organized economic landscape, the financial situation of ASADAs reflects the state of contemporary financial capitalism. ASADAs, like corporations and state agencies, operate under conditions of perennial debt, unbalanced accounts, and accounting innovations. But unlike those larger organizations, ASADAs also depend on the sheer commitment and volunteerism of community members. Plumbers, neighbors, and local committees contribute, out of a sense of duty, an immense amount of unpaid labor that never enters the organizations' accounting records.

Ana's comment about the need to pay for the infrastructure that delivers water to people's homes was met with caution. People with Ana's sensibility, particularly if they are women, are often considered too defiant, problematic. Instead of offering suggestions about how to deal with particular situations, people like Ana pose questions that do not seem to have apparent answers. They expand the parameters of discussion, maybe too much, bringing up the recalcitrance of the structures people want to, but seldom can, transform. Most of the time, people like Ana are dismissed as being too idealistic or radical.

But Ana's comment that morning was more pragmatic. Her invocation of God, water, and affordability was an honest and frank reminder to all of us that, first, bringing water to our homes does involve an unimaginable number of seemingly unrelated economic decisions that turn a godly gift into a worldly concern. Second, it reminded us that affordability continues to be a crucial concern for people as they go about their everyday lives. Ana was inviting those present at the workshop to reconsider carefully the economic implications of implementing a human right to water, to think about people's capacity to pay for it.

Her ideas disrupted what had been a fairly purified discussion hinging on the distinction between the "authentic" value of water and its financial value after commodification. References to the iconic Bolivian "water war" of the early 2000s served as a reminder of the perils of privatization and other ways of opening the door to the mercantilización of water, as we saw in chapter 1.[3] By introducing some friction in our conversation, Ana recast a discussion worthy of any philosophical salon as a concern with the nitty-gritty of billing options. Denaturalizing the monthly routines by which the ASADAs collect their income, Ana created space to wonder about why things were this way. How could a sense of affordability entangle God and Nature with water users and their unpaid bills? How were conflicting moral allegiances interlaced with practical demands to secure people's ability to take showers, cook, and have a glass of water when thirsty?

Not surprisingly, people at the workshop were drawn to Ana, myself included. During the coffee break, a group of us surrounded her while she spoke about the multiple dimensions of affordability, expanding on how the complexity of the issue could leave one paralyzed. She was clear that affordability could mean many things to many people. For a poor household, for example, affordability could mean deciding between paying their water or electricity bill. For a rich household, it could mean paying for an illegal connection to fill their swimming pool. For an ASADA, it could determine whether a plumber is paid to fix a broken pipe or is asked to do so as a personal favor. For large utilities, affordability could result in higher or lower delinquency rates. For regulators in ARESEP (the Public Service Regulation Authority), it could result in highlighting or downplaying the financial viability of ASADAs when regulators assess their methodologies. But recognizing such multiplicity of meanings was not enough for Ana. Just comparing different senses of affordability did not suffice. That was just a first step to survey a landscape, something like an act of setting a context so that a focus can emerge. What Ana saw as the necessary next step was identifying a point of entry into that multiplicity, an intervention that was not caught up by the conservative effects that documenting multiplicity can have as it overwhelms the senses with its excesses. Knowing this, she posed another question: What can we do if affordability means so many things at once? She then recommended, "I think that you should focus on prices, on how to keep them affordable. You should look into ARESEP and the prices they set for all of us."

I want to take Ana's injunction as the starting point of this ethnographic incursion into the technical twists that shape the affordability of water prices over time. Ana directed our attention to one specific place that morning: ARESEP, the regulatory agency responsible for setting the price all Costa Ricans pay for water. We were there in chapter 1 when we saw how regulators assess a utility's request for a price increase by interrogating their financial practices. Ana was signaling another dimension of regulatory work, noting how regulators also need to bring the subjects of rights, and not only utilities, into their calculative routines. In this chapter I show how, with an index, economists and statisticians record and reconstitute subjects in relation to the objects found in their households. This is how the affordability of a human right is produced and updated.

AFFORDABILITY AS PERCENTAGE

The affordability of public services has long been a concern for economic regulators in Costa Rica's public service regulatory agency. ARESEP regulators channel their preoccupation into experiments with different technical measures such as cross subsidies and tariff structures to make water cheaper. These measures are never as targeted or efficacious as they would want them to be; nevertheless, their attempts have helped expand water access to about 97 percent of Costa Rica's population. Their decisions have contributed to the creation of a state that shows how it "cares" for its population in this particular way (Foucault 2009; Mosse 2003). But in the 2000s, the work regulators do to keep water affordable went from being a national concern to an international benchmark, a quantified goal determined by the United Nations Development Programme (UNDP).

Intended to prevent price escalations such as the ones that followed the global water privatization wave that characterized the end of the twentieth century, in the twenty-first century affordability became a crucial component of the definition of a human right. In 2003, in what has come to be known in international law circles as Commentary 15, the Committee on Economic, Social and Cultural Rights of the office of the High Commissioner for Human Rights produced a definition of the right to water that would become normative. The Commentary states that "The human right to water entitles everyone to sufficient, safe, acceptable, physically accessible and *affordable* water for personal and domestic uses. An adequate amount of safe water is necessary to prevent death from dehydration, to reduce the risk of water-related disease and to provide for consumption, cooking, personal and domestic hygienic requirements" (Office of the High Commissioner 2003; emphasis added).

Of course, any of the adjectives used in the Commentary to specify the right to water can be subjected to a thorough analysis to understand its actual scope. But it is the question of what constitutes affordability that continues to garner considerable attention, in part because of its unstable meaning. To address the ambiguities implicit in the definition, the international establishment has produced a variety of guidelines, although one has become widely accepted despite its weak legal legitimacy. In a technical document disseminated widely across international networks of water-concerned professionals and activists, the United Nations Development Programme determined that if the human right to water was to be affordable, households should not spend more than 3 percent of their income to pay for it (United Nations Development Programme 2008).

When I learned about the UNDP's benchmark, I was surprised by its precision. Sofia, the ARESEP regulator we met in chapter 1, had explained that a percentage was not the best way to deal with a complex concept like affordability. Nevertheless, she considered the 3 percent figure helpful because it allowed people interested in human rights within ARESEP to have more focused discussions with colleagues and supervisors. The 3 percent figure helped concretize a general principle; it staked a numeric claim on a world that regulators can only engage through the tangible abstractness of their mathematical figures. Without a specific number, affordability could easily be brushed aside as a worthy but unworkable aspiration, something impossible to engage mathematically. It could become a general principle that everybody agrees upon but no one commits to in their daily work. With all its limitations, arbitrariness, and concrete vagueness, 3 percent was a number that ARESEP employees could engage with.

National statistics tell us that Costa Rican households use between 1 and 3 percent of their monthly income to pay for water. The households serviced by large utilities like AyA find themselves in a fixed payment regimen, meaning they receive their bills once a month, something that is now a metaphor since in 2012 an electronic system replaced paper bills. Once a bill is due, people no longer have a grace period to make their payment. If you fail to do so by the due date, the next day a private contractor stops by your home, opens the metal lid that covers your connection to the main line, and disconnects you. He "cuts your water" (*corta el agua*), people commonly say. There is no room for negotiation in this situation. To restore your connection, you can pay your bill at any of the small stores, pharmacies, or grocery stores now connected to the online payment system. One to three days later, depending on your neighborhood, the same contractor shows up to reconnect you to the water line. If you carefully read your next bill, you will see charges for the quantity of water used, as well as a "reconnection fee" plus the amount of interest charged based on how late the payment was. If you do not have a computer, or the patience to analyze the message, the details of these penalites remain invisible.

This strict policy was instituted by large utilities in the late 1990s as a way to discipline the population's payment practices. Up to that point, being late paying for your public services—water, electricity, telephone—was not cause for suspension. Households and businesses could be two, even three months late with their payment and continue to use those services. As part of the modernization of the public sector and the consolidation of the citizen into a consumer, public utilities and other state agencies adopted strict payment policies, resulting in a drastic transformation of people's relations with state agencies. Today, many of those relations look very similar to those based on transactions with private businesses.

The rural neighborhoods serviced by ASADAs have more flexibility in their billing and payment practices. In many of them, networks of informal subsidy allocation and lenience with late payments structure the rhythms and routines of monthly billing. Yet, in those areas, it is not uncommon for households to spend up to 10 percent of their income on water services. In Cocles, a small coastal community on the Caribbean coast, for example, I met Cristina, a mother of five and head of household who had recently lost her job at a nearby hotel and whose only income was a government subsidy of approximately $70 a month. Her monthly water bill was about $6, close to 8.5 percent of her income. But, as Alvaro, the fontanero with whom I was doing the rounds and reading people's meters told me (see chapter 1), the ASADA is very lenient with people like Cristina. "If you are a good person, you could never deny water. We know water is a human right and we respect that."

Regardless of whether it is an ASADA or a large utility like AyA that provides a household with water, ARESEP determines the prices people will pay and adjusts them cyclically. As Ana implied, sometimes it is possible to find a juncture, a particular location where certain people make decisions that make a big difference. One technical decision about the meaning of affordability can affect all sorts of practical questions about the material configuration of collective life. This technical decision has cascading consequences that activate certain genealogies and not others, creating certain preconditions for the future and not others. The price adjustments ARESEP makes are one such technical decision that has cascading consequences for whether households can live within their budgets, whether corporations decide to lobby the administration to get a new water extraction permit, whether a public school has enough money to add fruit to the meals they offer to their students, and an unimaginable number of other implications.

Fewer than twenty people worked in the Water and Environment Department (WED) of ARESEP when I began having conversations with them in 2008.[4] WED was rather small compared to the other three departments in the authority—telecommunications, energy, and transportation. Thinking that they had a much greater workload, experts from other departments were a bit dismissive of WED. They felt that regulating bus and taxi fares or oil and electricity prices was a "much more technically and quantitatively demanding" task than setting the price of water. This attitude became more extreme in years of oil-price roller-coasters, like 2008. That year, regulators received one petition after another to increase the price of the services most tightly connected to international oil prices: public transportation and gas. As we saw in chapter 1, each of those petitions triggers an administrative process that requires a lot of financial data collecting, mathematical analysis, and paper. As more petitions come in, the workload of regulators increases.

In 2008, ARESEP received only one water price increase request from AyA. During the previous three years the utility had not petitioned any increases, something very unusual. Thus, when they made their request they asked for an increase of close to 40 percent. AyA declared that without that increase, they would be bankrupt by the end of the year. That could be avoided, they argued, if their price increase was approved. With this petition, ARESEP's responsibility to keep water affordable was, once again, activated. Within WED they had developed a system so that each incoming petition was assigned to a lead expert who would be responsible for moving the process forward. That lead expert ultimately wrote up the technical recommendation that went to the head of ARESEP, a person whom technical personnel refer to as "the regulator." Ultimately, it was the regulator who signed the resolutions that all departments, including WED, drafted for him or her. That signature brought into legal existence the decision to accept, partially accept, or deny a utility's request.

Don Marcos was still in charge of WED during part of 2008. His position reported directly to the regulator. Don Marcos directed WED for more than thirteen years before he was relieved of his supervisory duties due to major organizational reforms. We often talked about what exactly regulation was all about. Once he explained to me that the job of a regulator is to act in the name of something else. He said, "We have to solve the absence of a market with a methodology. Because there are no natural water markets where suppliers compete for customers, our job is to produce the effect of competition with something else. We re-create the market effect for a public service with a methodology. In a way, we are the market." After finishing this statement, Don Marcos chuckled softly, as if acknowledging the strangeness of the responsibility of replicating the market effect with a mathematical pricing methodology. His chuckle was tied to a commonly held belief in Costa Rica. If something is "public," a good or service, it cannot be subject to what people imagine as market laws of supply and demand. The market dynamics that popular economic imaginaries see determining the fate of regular commodities are considered inappropriate for public goods.

In the absence of the competition implied by supply and demand ideologies, Don Marcos and his team are granted the legal authority to create an analogous effect. According to many of the theories they abide by, market dynamics have the effect of lowering prices and increasing quality. Even if personally they have questions about this established idea, the instruments they use have that belief embedded in their models, formulas, and principles. Therefore, WED regulators see themselves as doing two things. One is compensating for the market's shortcomings when it comes to setting the price and quality of public services. The other is producing the market effect through the design and supervision, one could even say care, of the social life of prices. This extraordinary dual responsibility has a clear name in regulatory theory: it is referred to as the surrogacy function of regulation (Jouravlev 2001). For Don Marcos and other regulators at ARESEP, this is not a metaphor; they see themselves as literal surrogates of the market while believing that water should not be regulated by the market in the first place.

SURROGACY IN THE FLOW OF TIME

The everyday work of surrogacy is not particularly spectacular. It requires a constant assessment of the traces left behind by history and the signs that anticipate a wealth of future events that regulators never get to witness. All that supervision, analysis, and decision making refers to events that occur outside of ARESEP's walls, beyond their immediate purview. What is left for them to work with are financial, accounting, and statistical statements that are taken as traces of past events and as projections of future happenings. In making everyday regulatory decisions about those traces and projections, WED follows two clearly determined procedures: ordinary and extraordinary price reviews. While theoretically these reviews can result in price increases or decreases, without exception, utilities use them to request increases. The technical specificities of those reviews turn what might otherwise seem mechanized bureaucratic routines into intense moments where the financial life of a human right is determined.

Extraordinary price reviews begin with a utility's request to reassess the components of the prices they charge due to major contextual changes—for example, environmental disasters, the adoption of new technical standards that require significant investments, and so on. Water providers do not experience major contextual changes of this kind often; nevertheless, they regularly request extraordinary price adjustments. Most often, they argue that unexpected costs are producing budget deficits that will ultimately result in bankruptcy. On most occasions, however, regulators determine that circumstances are not as serious as utilities claim them to be and do not warrant price increases of the magnitude that utilities request. But this does not mean that the utilities' petitions are flatly denied. On the contrary, when there are no special circumstances to justify an extraordinary price augmentation, regulators examine the effects of inflation on the "cost structure" of utilities and often grant some type of increase. They at least authorize petitioners to increase their prices by an amount equal to the inflation rate. Often this is the maximum amount regulators feel ethically authorized to grant.

The second type of review regulators perform is known as an ordinary price review, an annual reevaluation that is required by law. This review is designed to adjust prices according to factors that can be expected to vary regularly such as foreign exchange rates, oil prices, minimum wage regulations, and interest rates. Since regulators and other professionals in this field expect the cost of these factors to increase over time, they are predisposed to adjust prices upward. They calculate how much relevant costs have increased based on current data and use the inflation rate to supplement any gaps in information. The results of these ordinary reviews contribute to people's sense that fundamental services, and more generally the cost of living, do nothing but incessantly go up.

Both types of review, ordinary and extraordinary, are intensive exercises in temporal determination; they are opportunities for interpreting pasts and futures at once. Regulators interpret the past by requesting that the utility produce records of its recent history in the form of financial statements. When regulators review these records, they view them as historical evidence of two things: of the utility's due diligence and efficiency and of the utility's potential need to collect more income via higher prices. In their interpretations of the future, regulators rely on even fewer sources of information. Because future income and expenses have not occurred, regulators use the available records of the past and project them into the future. They create a future image of a utility's operations for a period of three to five years. These projections are traces of the future, notations of a time that regulators and water suppliers have a feel for but can never predict with absolute precision. These projections produced out of past records allow regulators to fold past and future into the present. They generate a speculative futurity that includes singular events such as, for instance, the negotiation of a new international loan or the purchase of major equipment. These kinds of future events can be anticipated through commitments from financial institutions or offers from machinery dealers. But the largest component of that speculative futurity consists of events that cannot be easily singled out. These happenings are a massive number of daily actions and transactions that make the flow of water through pipes possible—equipment maintenance, purchase of new parts and chemicals, salaries and benefits, electricity costs, communications expenses, public relations campaigns, gas for utility trucks, and so on. For a utility employing thousands of people, as AyA does, precisely anticipating those events is impossible. Thus, even though their connection to a financial history makes future projections seem precise, the sheer quantity of the actions that those projections claim to predict precludes any precise calculation. Rather, those projected actions and transactions can be better imagined as massive waves, enormous accumulations of dynamic events that continue in time and for which only mass groupings, traces in the form of budget lines or broad accounting categories, can be consulted. Within ARESEP, this futurity is regarded as an approximation, and a speculative one at that.

This folded temporality gives Don Marcos and the rest of the WED team a very sophisticated working relationship with time. They know that the world is in constant change, and that such fluidity changes the economic significance of the factors on which utilities depend to keep water of acceptable quality moving through pipes at the right pressure. Thus, to be able to rely on accounting records and projections, despite their uncertainty, regulators require an instrument with the capacity to simultaneously adjust pasts and remain sensitive to futures. They need a tool with the capacity to counteract the sense of indeterminacy that their calculations always possess. They need a device with sufficient ontological and moral legitimacy to adjust a price without breaking its coherence with the 3 percent humanitarian ideal.

The numeric instrument deemed capable of performing all that technical and moral work is the inflation rate, that macroeconomic figure that holds the capacity to accelerate or decelerate economies, classify a nation as dynamic or stagnant, and explode or implode the number of calculations people perform daily as the value of their money deflates or inflates. The inflation rate is a cornerstone of economic thought and practice globally. When other economic concepts are deemed untrustworthy, people turn to inflation as a solid indicator to base decisions on. Take, for instance, discussions over the situation of the U.S. economy at the end of 2017. After questioning the reliability of concepts such as "natural rate of unemployment or neutral real rate of interest," a former member of the board of the Federal Reserve affirmed that the only trustworthy source of information was inflation (Fleming 2017).

The inflation rate has a very precise function in the methodology regulators at ARESEP use. If no other justification can be found to increase the price of water service provision, but they still believe the companies are entitled to an adjustment in their price, they increase it by the inflation rate of the previous year or the projected rate over the next year. Inflation is the mathematical factor regulators, and economists more broadly, trust with the capacity to make past traces current and future occurrences present.

In a strict sense, inflation is the measure of decreases in the value of money due to increases in the prices of goods and services. The more inflation there is, the fewer commodities a given quantity of money can purchase. The enormous trust regulators place in the inflation rate stems from two sources. First, there is the power of a figure that has been agreed to be a fundamental economic indicator. For decades, textbooks, policy makers, and central banks have used the inflation rate to represent the health of a national economy. Over time, the power of this figure has become taken for granted, creating a sense of inertia that privileges its use across specialized groups, including regulators and many everyday citizens. Second, the regulatory procedures, and more generally the rationalities and methodologies that go into the inflation rate, imbue it with a kind of specificity and concreteness. By using the inflation rate in their calculation, regulators make an analytic leap. They use the inflation rate as a mechanism to incorporate into their pricing formula all those things that are counted in the production of the rate itself—the intimacy of households, the things that are brought into them, the services people choose. All of these very household-specific and highly intimate factors are intertwined through a calculative process I describe below and which economists believe has the power to channel all these hyper-context-specific relations.

As we can see, with the act of taking the financial statements and projected cash flows of utilities and multiplying them by the inflation rate, regulators are doing a lot more than merely augmenting their magnitude. They are in practice responding to Ana's question about how affordability is determined in the flow of time. They are also turning the ideological and methodological meaning of inflation into an unexpected space for the elucidation of humanitarian reason. Regulators' reliance on inflation, and the numerical index on which it depends, expands the scope of humanitarian logics usually enacted through courts and emergency aid. The inflation rate brings these logics into the everyday action of accounting and financial modeling. In this surprising way, the inflation rate comes to matter as regulators attempt to bring about the difference a human right can make in people's lives.

MATERIAL-SEMIOTIC MULTIPLICATION: FROM GOLD TO COMMODITIES

Puzzled by the fact that money can lose and gain value without seeming to change its nature, economists have developed multiple techniques to track and predict those variations and the effects they have on employment, investment, and growth (Neiburg 2006). The inflation rate is without a doubt the most important tool for tracking those variations; it is so central to economic life that in Costa Rica landlords use it to update the rents they charge, national authorities compare the adequacy of minimum wage to it, and regulators take it as the only irrefutable justification for updating the prices of all public services including electricity and water. Not even the threat of bankruptcy enjoys such power in their analyses. Because of its impact on so many dimensions of economic life, the inflation rate is a "machinery of relating" (Holmes 2009: 6), a figure capable of "articulating policy in relation to both the distinctive and shared circumstances of individuals . . . who are continually modeling and transacting economic relations" (6). And while its role in turbulent political and economic life has been carefully examined (Holmes 2009; Roitman 2005), the power of inflation to shape social life in unexceptional times remains for the most part invisible.

Broadly, the inflation rate is a percentage that tracks the fluctuation of the level of prices on an annual basis. But from the point of view of its mathematic production, the inflation rate is nothing more than the recalculation of the consumer price index (CPI). The CPI is a figure that quantifies the relation between money (prices) and things (commodities) at two points in time, usually from one month to the next. The CPI is the vehicle for two important pieces of information. First, it includes a particular group of commodities, referred to as the consumption basket. This basket comes to signify an idealized household; it is a set of commodities that stands for what one can find in a home that represents the "Costa Rican household" as a general category. The other piece of information the CPI carries is a historical record of change. The final numeric form of the CPI is a percentage that indicates how much the prices of the goods in that basket have changed from one month to the next. The CPI carries this information, a collection of goods and a change in time, into the heart of the inflation rate, which in very simple terms is nothing more than the summation of the findings of the CPI over the period of a year.

Originally inspired by the labor theory of value, the CPI can be taken as a historical account that stands for the "public consideration of what is involved in making a living" (Guyer 2013: 13). The CPI is an intensely historical articulation of the things people purchase, their presence within the household, and the quantity of money people have available to satisfy both their needs and desires. For economists, the combination of all of that information into a single figure is very powerful. Not surprisingly, the CPI has been used to answer moral questions about what is entailed in making an acceptable living, what is the place of one nation—as an economic entity—in comparison to other nations, and how healthy or weak both figurations, the nation and people's capacity to make a living, are (Neiburg 2010).

In Euro-America, one of the first documented uses of a construction like the CPI goes back to 1707 with the bishop of Fleetwood's publication of a book titled *Chronicum Preciousum* (Kendall 1969). In that book, alongside his concerns with miracles and sermons, the bishop developed his ideas about the changing value of money, gold, and silver. His analysis was a response to a very practical moral problem presented to him. As it turned out, a college founded in the 1400s required its fellows to take an oath that if they accumulated an estate of more than £5 they would vacate the college premises. In the early 1700s, a fellow of that college was worried about the current moral value of the oath. He asked the bishop whether he thought it reasonable to take such an oath considering how the value of money had fallen since the 1400s. The bishop embarked on a study of the previous three hundred years by comparing how much "corn, meat, drink and cloth," items he thought were the only necessities of an academic, could be bought with £5. He concluded that £5 in the 1400s were equivalent to between £25 and £35 in the 1700s, and recommended that the fellows' oath be updated to reflect that value.

Despite their practical importance, explorations like the bishop's were deemed comparatively unworthy endeavors in relation to philosophical examinations of the metaphysical nature of value and its transformations. The lower status given to those wonderings was due to their closeness to the concerns of the domestic space and its worldly nature, as opposed to the public sphere of intellectual work and to questions of theological importance, which were considered of "higher order." Authors doing those kinds of mundane mathematical musings feared appearing "to descend below the dignity of philosophy, in such oeconomical researches" (Evelyn 1798: 176) and often agonized about how their "voluntary pursuits should be directed to higher purposes than worldly objects, whatever temporary importance may attend them" (Arthur Young, cited in Kendall 1969: 5). And yet, despite being described as a "descent" to the level of worldly objects, the preoccupation with how many things money can buy has survived for centuries. Today, the inflation rate and the CPI are central indicators not only of the economic health of the nation, but also, as it turns out, of the everyday life of humanitarian concerns.[5]

My own joyful "descent" from the philosophical exploration of the meaning of humanitarianism into water as a "worldly object" also took me to the CPI. The relation between people, prices, and things quickly became central as I encountered a new iteration of the consumer price index developed by Etienne Laspeyres in the nineteenth century. Sofia, the regulator we met in the previous chapter, directed me to this index when she explained the significance of inflation in their price reviews. I first saw its mathematical expression in a booklet produced by Costa Rica's Statistics Institute to disseminate information about the index and how it is produced.

$$I_G^t = \frac{\sum_{g=1}^{n} = I_g^{t*W}{}_g}{W_g}$$

where

I_G^t = General index of the month being monitored

I_g^t = Index of group to which an item belongs in the month being monitored and in relation to the reference period

w_g = Group weighting

w_G = 100

Equation 2.1. Laspeyeres's index as used to calculate Costa Rica's monthly CPI.

Etienne Laspeyres was a Dutch economist and statistics scholar who in the 1800s, unlike others at his time, was convinced that the problem of understanding inflation demanded a study not only of changes in the availability of gold but of the changes in the prices of everyday goods. He argued that changes in the value of money were caused by peculiarities in people's material needs and desires and did not exclusively reflect the value of gold (von der Lippe 2012: 338). To show the relevance of everyday things to larger economic questions, Laspeyres considered the quantity of objects people purchased, their importance in relation to the total number of goods and services a household acquired, and the proportion of money they used for acquiring them. He thought relationally, putting goods side by side, and he also thought proportionally, comparing the cost of one good to total income. By following Laspeyres's formula we learn, for instance, that in 2015 people in Costa Rica used more of their income to pay for rice than to pay for a belt, and that both of those items take more of a family's income than a mattress (Instituto Nacional de Estadística y Censos 2016). In other words, Laspeyres realized that not all things are created equal and that the quantity purchased and the proportion of a household's budget allocated for the consumption of an item were all important indicators of the different significance that objects have in people's lives. Laspeyres's original formulation has inspired virtually all the equations used around the world to calculate a CPI.

INDICES

As economy-making devices, "index numbers [such as the CPI] encapsulate the entire dynamics and circularity of relations between economic theory and economic cultures" (Neiburg 2006: 615). While they are described as indicators of a preexisting reality, economic indices are performative instruments, devices that bring into existence the reality they claim to merely describe (Callon 1998; Mitchell 2005). In their use, their proclaimed function as trackers of economic practices (consumption ones in this case) turns indices into devices with the capacity to create connections and separations that differ from those originally foreseen by their creators. Laspeyres, for instance, could not even conceive of a connection between his index and the quantification of the affordability of a human right. Part of the expansive power that allows indices to participate in such diverse world events comes from their semiotic capacity to stretch across time and space. That capacity to stand for multiple events at multiple moments in time is so broad that some indices are interpreted as being the very embodiment of markets and even of the "economy" as a whole (De Goede 2005). Interestingly, that all-encompassing capacity is not concealed in any way. To the contrary, people are drawn to indices precisely because they openly present themselves as multiplicities, as relations between other entities, as combinations of multiple things and world-events. They are never taken for singularities. For their users, the power of indices resides in their composite character. That is one of the main reasons why regulators and other economic actors like to work with indices such as the CPI, because they are self-evident combinations of other things that give a sense of concreteness and context specificity.

Numeric indices accomplish this semiotic relationality by way of their multiplicative logic. A multiplicative number has the ability to embody the relative weight of each of its factors and pass it on to the final calculation. It is a trace of relations that are not homogenous. In the CPI, each component is weighted differently, and for that reason its relation with the final number has a particular, rather than a standardized, significance. For instance, in Costa Rica's 2015 CPI, "water services" accounted for 1.41 percent of the cost of the consumption basket, while "drill" accounted for 0.05 percent, "birth control" for 0.11 percent, and "pizza" for 0.35 percent (Instituto Nacional de Estadística y Censos 2016). Sofia and Don Marcos do not think of those difference in terms of their concrete magnitude. They understand them as relations of different intensity, as textured and concrete representations of the financial aspects of people's everyday lives. In that sense, what matters about the CPI is the proportional relations between the cost of a specific object and the set of all other objects being purchased, as represented by the consumer basket. Don Marcos and Sofia take these proportions as evidence of the relation between people's everyday needs and their purchasing decisions. These proportional relations are the specific moments at which the concrete needs of life—eating, drinking, caring for others, having the luxury of enjoying leisurely goods—meet specific financial conditions of possibility. For that reason, as the relations between water, income, and the other commodities change in time, regulators rely on the CPI as the only legitimate proxy for the changing conditions of the Costa Rican household that they can use in their calculations. Thanks to that sense of specificity, the CPI and the inflation rate work as depictions of some empirical realm that justifies their numeric association into something as fundamental to life as water.

THE SLOW INDEXICAL DISSIPATION OF SUBJECTS

By 2016, Costa Rican statistical and economic agencies had produced seven iterations of the CPI. The names given to each effort provide a historical record of the changing economic and political rationales behind them. The very first CPI on record was calculated in 1936, and was named "the cost of life index." The next one, in 1952, was said to measure consumption "by middle-class consumers and working-class citizens." In 1964 and 1975, the index was described as a measure of the "purchasing power of middle- and low-income consumers." And in 2004, the Statistics Institute began describing its calculation of the CPI as a "generic consumer index," without any reference to people or their economic class. Thus, at the beginning of the twenty-first century the CPI became a sign of an abstract consumer. Cost of living, working citizens, and middle- and low-income consumers were replaced by an unmarked collective of commodities undergirded by a human entity whose existence is only asserted by the purchases it makes. But these nominal changes were more than changes in labels. They were based on methodological innovations in the statistical parameters used to define the household whose things and income the CPI measured.[6] And while these changes were statistically explained as a way to produce a more accurate image of the country, conceptually the shifts mark an important transition from humans to things, the gradual numeric effacing of the human as a parameter defining a household and ultimately, for my purposes here, a human right.

1930s: cost of life →
1950s: middle-class consumer and working-class citizen →
1960s and 1970s: middle- and low-income consumers →
2000s: unmarked consumer →
unnamed and unmarked human[7]

Until 2004, the CPI statistically included households with between two and twelve residents. Households with more or fewer residents were excluded as statistical outliers. That 2004 calculation of the CPI was the last to include in its defining parameters any explicit reference to number of individuals occupying a household. From that point onward, the mathematical design of the surveys on which the CPI is based included households of any size, regardless of the number or economic situation of the individuals in them. The only information incorporated into the calculation was goods and services consumed; people dissipated as statistical figures defining what a household is. From one point of view, this seems a welcome shift. No longer was the state deciding what counts as a household using traditional and heteronormative ideas of what kinship is. From another point of view, however, this shift also marks a change in how relevant people are to the definition of the household in comparison to the goods and services they purchase. While this shift from people to purchases might seem a small technical change, it is more than that. It is the statistical culmination of a trend for which the specific conditions of kin and class relations, with their historical particularity, are pushed into the background. Instead, we see the statistical emergence of the person in the household as an unmarked potential consumer. In this scenario, small human collectives, kin groups, become equal to their potential purchases. The household as an important unit of sociality is transformed into an empty signifier—no assumptions of a nuclear family, a heterosexual reproductive nucleus, or an extended system of relations of care are necessary. The household becomes a statistically determined collection of things and services, the accumulation of the material traces left by transactions and, only by association, a trace of the people who happen to be connected to those things and services.

In this new landscape, the objects and services included in the consumption basket are the parameters that define what a household is. This shift is expressed as follows in the statistical method currently used to calculate the CPI. To be part of the basket the CPI traces, a good or service must fulfill at least one of two requirements. It must consume 0.5 percent or more of the monthly expenditures of a household, and/or it must be consumed by at least 5 percent of the surveyed households. In the last twenty years, the collection of items put together by applying those rules has grown from 264 items in 1995 to 315 in 2015. In tables 2.1 and 2.2, we can see a selected list of items that were added into and removed from the consumption basket in 2004 and 2016. In the 2000s, households introduced into their homes more "expensive" objects, changing in the process the very "nature" of this imaginary Costa Rican household. Things like candles, liquid floor wax, beets, and pantyhose were no longer part of the basket. Instead, new goods and services such as cable TV, internet service, ultrasound, yogurt, car wash, and cell phone service were included in the basket. Further, in 2015, "bottled water" made it into the CPI, for the first time.

Each of these 315 items deserves a cultural history of its own. For instance, the changing preferences among younger generations of women toward wearing pants instead of skirts or even the increasing desire for "exposed women's skin" might explain the disappearance of pantyhose from the basket. Changing preferences in the Costa Rican palate, maybe the growing interest in fast food or changes in the allocation of subsidies for farmers, could explain why beets disappeared. But individually, each of the hundreds of objects and services in the index has only a minor impact on the CPI and ultimately on water affordability. For purposes of the calculation of the price of water, their individual significance is subordinated to their mere inclusion in the index. What matters is not a single item, but the articulation of 315 of them as a unique set of proportional relations that change month to month. These idiosyncratic relations, translated into the CPI and later into inflation, have the power to shape the future of a utility and of water when such a future cannot be accounted for with precision.

But there is another element that confounds the significance of the CPI as a set of concrete and context-specific relations. Because of its statistical composition, there is no single household where one could find the 315 items the CPI accounts for. The CPI is only a statistical image of a household that is in reality distributed throughout the more than 7,000 households surveyed to determine the contents of the consumption basket. The CPI household is an abstraction designed to provide specificity and concreteness. It is a figure regulators use to create a context, to explicitly link

Table 2.1 Changes in the 2004 Consumer Price Index

	ITEMS EXCLUDED		ITEMS INCLUDED
1	Liver	1	Baked goods
2	Sardines	2	Hotel services
3	Fresh milk	3	Ink cartridge
4	Processed cheese	4	Pork ribs
5	Green celery	5	Rice cooker
6	Green pumpkin	6	Magnetic backup unit
7	Cauliflower	7	Chicken wings
8	Beet	8	Iron
9	Sweet potato	9	Video game
10	Unprocessed sugar	10	Breaded chicken
11	Dry stock cubes	11	Coffee maker
12	Powder cacao	12	Cable TV
13	Achiote	13	Sausages
14	Pepper	14	Screwdriver
15	Panty hose	15	Touristic packages
16	Women's dress	16	Condensed milk
17	Skirt	17	Body lotion
18	Short sleeve shirt for child	18	Cream cheese
19	Women's blouse	19	Internet service
20	Boy brief	20	Movie rental
21	Girl panty	21	Yogurt
22	Girl's dress	22	Cloth softener
23	Baby plastic pant	23	Book
24	Baby sock	24	Avocado
25	Boots	25	Trash bags
26	Shoe hill sole exchange	26	Veterinary services
27	Shoe sole exchange	27	Mattress
28	Cotton	28	Foreign language classes
29	Thin cotton fabric	29	Grapes
30	Linen	30	Air freshener
31	Silk	31	Computer literacy course
32	Blanket	32	Sweet corn
33	Electric buffer	33	High blood pressure medication
34	Blender	34	Dictionary
35	Ceramic tableware	35	Mushrooms
36	Liquid shoe polish	36	Vitamins
37	Paste shoe polish	37	White paper

Table 2.1 Changes in the 2004 Consumer Price Index *continued*

ITEMS EXCLUDED		ITEMS INCLUDED	
38	Light bulb	38	Peas
39	Scouring pad	39	Anti-allergy medication
40	Candle	40	Photocopies
41	Liquid floor wax	41	Refried beans
42	Solid floor wax	42	Cough medicine
44	Matches	44	Chewing gum
45	Broom	45	Ultrasound
46	Soap bar (for washing clothes)	46	Photocopy
47	Effervescent analgesic	47	Gallo Pinto (rice and beans dish)
48	Antacid	48	Stock
49	Fortifying food supplement	49	X-rays
50	Alcohol	50	Photographic camera
51	Pain relief icy cream	51	Lawyer fees
52	Tire alignment	52	Watermelon
53	Tire balancing	53	Funerary services
54	Tape recorder	54	Ironing table
55	Tricycle	55	Printer
56	Photographic film	56	CD
57	Film developing service	57	Party service
58	Hair spray	58	Furniture repair

their work with the household of Costa Rican water consumers, despite its inexistence in space and time as a brick-and-mortar household. This is no surprise to regulators, nor do they make apologies for this fact. In ARESEP they recognize this as a trade-off, one of the insurmountable constraints they confront every day to keep open space for humanitarian affordability within their calculative routines until they can find a better way to enact the difference they want to make in the world. The CPI is the closest regulators can get to mathematically representing the citizen as they adjust the price of water to the passing of time.

As we can see, in the process of selecting the goods and services in the consumption basket, the Statistics Institute that calculates the CPI does much more than track changes in the value of money. It also produces an image of the material form of the household, a literal objectification of kin relations through prevailing economic and statistical orthodoxies. For national economic policymakers, the basket of goods and services operates as

Table 2.2 Changes in the 2016 Consumer Price Index

	ITEMS EXCLUDED		ITEMS INCLUDED
1	Whole fish	1	Pre-prepared rice
2	Mushrooms	2	Insurance
3	Peas	3	Snacks or appetizers
4	Chewing gum	4	Financial services
5	Fruit sauce	5	Engine repair
6	Women's socks	6	Hair coloring
7	Dry cleaning	7	Hospital fees
8	Cement	8	Pre-prepared pasta
9	Glass	9	Soups
10	Masonry service	10	Women's dress
11	Computer desk	11	Natural flowers
12	Furniture repair	12	Surgery
13	Ironing board	13	Women's bag
14	Sheets	14	Cake
15	Microwave oven	15	Zinc sheet
16	Pressure cooker	16	Served fish filet
17	Hammer	17	Motorcycle
18	Pliers	18	Served dessert
19	Rinsing aid	19	Diabetes medication
20	Cough medicine	20	Served salad
21	Car wax	21	Gastritis medication
22	Car clutch	22	Shift-stick repair
23	Car battery	23	Medication for nerves
24	DVD player	24	Shock absorbers
25	Photographic camera	25	Bottled water
26	Printer	26	Drill
27	Movie rental	27	Birth control
28	Magazine	28	Mobile electronic device
29	Dictionary	29	Kid's coordinate dress
30	Lipstick	30	Light bulb
31	Nail polish	31	Passport fees

a summary of the lifestyle of any household in Costa Rica, even if in reality most households are not able to afford it. It is an image that is normative and contextual at once. It is implicitly normative about the material foundation of the nation, the concrete meaning of a "Costa Rican household." It is contextual to the extent that, for example, it links the 3 percent affordability benchmark set by the UN with the "specificities" of households in Costa Rica.

The kinds of relations between people and objects that the CPI stands for are not new to anthropology. Since the early days of anthropological inquiry, scholars have turned their attention to the role objects play in shaping our (non)humanity (Evans-Pritchard 1940; Malinowski 1935; Mauss 1967). This literature has taught us to not misrecognize objects as passive receptacles of meaning, and instead see them as active participants in our world-making endeavors (Henare, Holbraad, and Wastell 2007; Latour 2005; Pottage 2001). What is remarkable in the case of the human right to water is how things, in the form of commodities, not only co-inhabit the world but have come to numerically stand for humans when we locate them in their households. Further, these relations between people and things are used to numerically qualify a fundamental right. This displacement of the subject by objects happens not in "alternative ontologies" but at the very core of a liberal society. Here, the subject of rights is mathematically rendered by the things we find in our household, and further, by gendered fashion trends, farming subsidies, and myriad sets of relations underpinning people's choices about the things and services they buy.[8] In an interesting statistical twist, objects push humans to the background as regulators attempt to keep water prices as affordable as possible. It is as if in the world where people try to make the human right to water into something concrete, things have emerged as a primordial and reliable substrate to account for the humanitarian subject. Things have become trustworthy, countable, and retrievable evidence of our existence. Instead of thinking about the human rights of subjects, we might have to start thinking about the rights of human objects.

This development seems even more odd when we remember that this is a world where people understand themselves as individuals relating to other individuals, as subjects commanding over the passive things that surround them. In this context, making the household of objects determine the affordability of a human right is an interesting transgression of secular separations of subjects and objects. Although not intentionally promoted by regulators or international human rights officials, this transgression is nevertheless statistically produced—it is made possible by the operations behind the CPI.

One should not think of this transgression as an exclusively Costa Rican development, though. Inflation and the CPI are used worldwide to make all sorts of decisions about collectives. They are mobilized to evaluate debates over the morality of the economy, the convenience of increasing interest rates, the magnitude of the minimum wage, the legal yearly increase in rent, the risk factor of a construction loan, and many others. Thus, by asking how and where the CPI is used to settle collective concerns about rights that take the form of a technical economic decision, we can see the ways in which objects in households, statistically aggregated into a mathematical index, have become a major indicator of our humanity. It is not that, as many critiques of capitalism claim, these economic indicators rob us of our humanity. Rather, what they do is change how we count as humans. They imbue the rights that make us human with commodities all the way down. Understanding how this happens sheds light on how economic logics, humanitarian and humanist commitments, and material ontologies coalesce into the numeric creation of the preconditions for the future.

HUMANITARIAN HOUSEHOLDS AND THEIR THINGS

The associations and creative forms of reasoning that I have followed here happen across a variety of centers of calculation. They are part of a formation akin to what, in his analysis of climate change, Paul Edwards (2010: 8) calls a vast machine: a sociotechnical system that collects data, models processes, tests theories, and ultimately generates widely shared understandings. There are many such vast machines; one of the most familiar to us is the census-making international machinery with its historically multiple objectives of creating commensurable units, making colonial control possible, and inventing the population. In Costa Rica, the raw data for the CPI is collected by the same entity responsible for the national census: the National Institute of Statistics and Census, or Instituto Nacional de Estadística y Censos (INEC).[9] This institute belongs to a network of similar institutions throughout Latin America, and really throughout the world, that not only produce statistical data but also, and importantly, help disseminate and regularly "modernize" statistical techniques. These statistical centers are the original "Big Data" collectors whose categories and counting techniques continue to have quiet but far-reaching consequences.

In 2012, Costa Rica's statistical machinery almost halted. Due to a conflict with the central budget authority, the positions of 462 of its public servants were nearly lost. When the director of INEC went public to gather support for his agency, he remarked that "a country without statistics cannot make good decisions." His comment was an affirmation of the positivist legacy of the elites of nineteenth-century Latin America who promoted political reforms inspired by their experiences of being educated in Europe. Today, INEC faces many problems. One of its major difficulties is that once it hires statisticians and trains them for years, they often move to private corporations in search of better salaries. To address this situation, and to stabilize its professional cadres, INEC has been working toward modernization, both attempting to increase the salaries it pays and launching new statistical projects to satisfy the increased "statistical demand" of the country. And it has certainly done so. For this research I requested data for the three most recent recalculations of the CPI, and the data was sent to me electronically in less than two days, likely a record for Costa Rica's slow-moving bureaucracy.

Rolando, a member of the permanent statistics office of INEC, is part of the team that every ten years conducts the Encuesta Nacional de Ingresos y Gastos de los Hogares (National Survey of Household Income and Expenditure), commonly referred to as *la Encuesta de Hogares*. INEC produces a variety of outreach materials to inform the citizenry that the Encuesta is about to take place and that it is their obligation as residents of the country to provide the information INEC requests. These flyers are distributed in supermarkets, in gated communities, and in other public spaces (figure 2.1). They target class and gender differentially and are accompanied by TV ads, newspaper communiqués, and information capsules on the radio.

This survey identifies the items included in the consumption basket and the CPI. No one is more aware than Rolando of the methodological changes that the CPI has undergone. And yet he continues to see the Encuesta de Hogares as a human-centered effort. One afternoon, as we discussed how the Encuesta is organized, he described the process by referring to the encounter between the people conducting the survey and the residents of a household. Out of our conversation and a short informational video on YouTube Rolando recommended, I put together the following image of the

Figure 2.1 Metamorphoses of the human in relation to the household and the CPI.

routine data collection visits behind the making of the object-based statistical household:

Door knock. Door opens. *Buenos días Señora* (Good morning, ma'am). I am from INEC. We are doing a survey; may I speak with you for a moment? While waiting for an answer, the surveyor imagines what things might be inside the house. How long will this survey take? Secretly she hopes there is not much in the house. She considers closets, pantries, bookcases, appliance stands. All sorts of structures that hold, protect, or hide from view the numerous things that inhabit this home along with its human and nonhuman dwellers. In this first encounter the surveyor is somewhat uncomfortable, although the more interviews she does, the more mechanical the initial exchange becomes. She has gone through training, mostly about how to complete the questionnaire appropriately, how to record answers, and where to hold the surveys. She is between twenty and thirty years old, without a stable job, and is about to finish her university degree in sociology. She responded to a newspaper ad calling for *encuestadores* (poll takers) to be temporarily hired to collect the data for the Encuesta de Hogares. Most times, but not all, the people who open the door are cautious women, elderly persons, or domestic workers.

Rolando reminded me that increasing rates of property crime have turned what used to be warm welcomes into quick assessments of the credibility and possible dangerousness of the person knocking on the door. Is she really from the institute? Is that ID real? Are there other people around her? While initially concerned with security, residents usually soften and most times invite the surveyor to come in and sit in the living room or kitchen and offer her something to drink—maybe *un fresquito* (a fruit juice), coffee, or at least *un vasito de agua* (a glass of water).[10] This scene is multiplied by the thousands every ten years when INEC conducts the Encuesta.

In the same way that Don Marcos visualizes concrete households paying water bills as they use the inflation rate to update their prices, with this hypothetical image, Rolando wanted to imagine the CPI calculation through the specific conversations and handshakes from which their statistical information is drawn. This link between statistical data and the concreteness of the day-to-day human interactions that make the Encuesta happen is a way of keeping the relation between people and things active, a way of entangling that which numerically they will disentangle. In other words, what in people's imaginary is an intricate relation between people and things, in their statistical decisions is the slow erasure of the specificities of the human—for example, gender, class, or kin relations—until she becomes a statistical nonissue.

This oscillation between that which they numerically claim and that which they visualize as the justification of their work is symptomatic of a tension between numbers as abstractions and counting as place-specific practices. Regulators and statisticians often confront this tension when everyday people compare their personal experiences to the economic figures that INEC and ARESEP present to the public. Brian Rotman (1997) argues that this tension is the result of the implicit numerical assumptions of Western mathematics. He argues that the material reality of counting does not allow for the abstracted infinity people presume is intrinsic to numbers. That abstract infinity is only possible in a Platonic tradition where numbers are taken as "an already existent, infinitely extended series of objects, each different from its neighbor by an identical unit" (36). But counting things in the world is not a process of identifying infinite and self-evident units. It is a social practice that requires different numerical logics according to the particular context where it unfolds. As Helen Verran (2001) notes, numbers and counting are practical accomplishments "embodied in collective goings-on in specific times and places" (220). Don Marcos and Rolando conduct their everyday work by interlacing these two logics: an abstract infinitude of Platonic numbering and place-specific regimes of counting and calculating.

Reflecting on the historical antecedents of the calculation techniques they used at the moment, Don Marcos, who was known among his coworkers for his good humor and storytelling abilities, recalled an anecdote from the 1970s when he worked for the regulatory body that preceded ARESEP. The story is another instance of the coexistence of these two forms of numeration, Platonic-abstract and place-specific. At the time, any price increase in public services had to be approved by the country's president. During the last government of José Figueres (1970–74), the president who in his first term in the 1940s nationalized banks, abolished the army, recognized the voting rights of women and black citizens, and also persecuted communists, there was a request for a substantial water price increase. Don Marcos couldn't remember the exact numbers, but he narrated the story of how they went to see Don Pepe, the affectionate name people still use to refer to Figueres. Don Marcos prefaced the story by saying, "The man was very funny and a little *atarantado* [a combination of amusing and inattentive]." After listening to all the technicalities of the request to increase water prices the president responded, "Very well, boys, but tell me something, how much is a beer these days?" Don Marcos and his colleague looked at each other and tentatively answered, "x amount." To which Figueres responded, "How many beers fit in a cubic meter and how much would that cost?" They did a quick calculation and told him how much it would be. The president then asked, "And how much are you asking to increase water by?" After they replied, Don Pepe thought for a moment and then said, "Okay! One hundred pesos, versus one thousand that the cubic meter of beer is worth? No problem. Go ahead with the increase."

Besides its amusing tone, Don Marcos's anecdote provides a broader historical context to the calculative relations between people and commodities, and their place in the definition of collective life. It shows how affordability has for a long time been about the place of water in relation to other purchased things. This is the kind of lateral association people make when they compare water to other things they pay for. It is also the kind of association that embeds the idea of a human right to water with the physical objects that make our material-semiotic lives livable. When used in technocratic spaces such as ARESEP, the CPI and the inflation rate provide a powerful answer to fundamental moral and political questions about how universal rights acquire concrete forms. In this case we see how changes in the CPI diminish the statistical prominence of the human in the definition of the household that ARESEP uses to make their economic measures contextually specific. Where the state used to care about workers, lower- and middle-class citizens, and even consumers, we now find an invisible person, an unclassifiable potential purchaser of goods and services. Proportionally to the fading of the historically and class-specific human, we can trace the ascendancy of purchased things and services, all the way to the definition of a human right to water.

In 2014, Nancy Fraser posed the question of whether societies can be commodities all the way down. Her question was framed by the environmental and economic crises that she sees unfolding globally. In this case we might ask a parallel question, one that Sofia, Rolando, and Don Marcos struggle with every day: Can human rights be commodities all the way down? In the case of the human right to water, it very well might have been statistically made so, if only provisionally, until the next calculation happens, the next methodological innovation is developed, and the next international benchmark is adopted.

CONCLUSION

Putting in place numerical and statistical structures for responsibility for the future, in this case the future affordability of water, requires regulators to contextualize a price in a household despite the inadequacy of their instruments for doing so appropriately. Here the CPI, in relation to the 3 percent benchmark, is a placeholder to create a distinction between a humanitarian and a nonhumanitarian price. The CPI is a device that creates a distinction; it makes a difference in a field fraught with logistical challenges and political opposition. Despite the acceptance by most everyone in Costa Rica that one should pay for water, the precise task of translating the human right to water into a very specific price unleashes all sorts of questions and doubts. Being a surrogate for the market, when you believe the market should not be dictating the price of a human right, requires this kind of inventive technique to make a difference in a world that constantly pressures people to keep things as they are.

Every price adjustment regulators perform is a temporary accomplishment followed by a new demand to incorporate new ideas and methodologies in their work and ultimately to adjust their prices according to the most recent inflation calculations. Thus, regulators know that as soon as they adjust a price, time is already creating the need for a new adjustment, for a new calculation, which is ultimately a new opportunity to create another ephemeral distinction between the human right to water and its commodification.

By tracing the calculations regulators make to try to keep water affordable, I ended with the CPI at the center of my analysis. The CPI is a number that accounts for the things and objects found in a statistical depiction of a Costa Rican household. The capacity of that abstracted household to purchase objects and services is transmorphed into an inflation rate that adjusts the periodic efforts of regulators to make water affordable for all, especially for the poor. Regulators acknowledge the fluctuations of purchasing practices and income among Costa Rican households and strive to align their prices to those fluctuations. Over time, the CPI and the inflation rate slowly become one of the last humanitarian devices regulators can use as their decisions become more and more embedded in financial logics.

But regulators' reliance on the CPI has further implications. As I have shown, the calculation of the CPI entangles things, persons, and water as a gift from God and Nature, challenging the purity of any separation between a right and a commodity. The CPI recasts the separation between persons and things, a fundamental distinction to liberalism, by making commodities the units that quantitatively shape the humanitarian character of subjects and their right to water. Through these extended connections, the CPI gives water an extended materiality that is much more than pipes, valves, and H_2O. Water, in its social life as a right, is a numeric choreography of familiar hydraulic infrastructures along with TVs, cooking implements, computer literacy lessons, bottled water, and the more than 300 items currently tracked by the CPI.

Thinking with this index moves us away from individual subjects, if we take them as self-evident units of rights or as the self-evident ethnographic "objects" that need to be traced. Through the history I have traced, the individual—that cherished subject of rights—went from being an entity with particularly classed histories and existing in a small collective that we call a household, to being a distributed entity only recognizable mathematically through the goods and services she has collectively purchased. In this household, the individual is dissipated into her consumption practices. In their thinking, people like Sofia, Don Marcos, and Rolando are avowed humanitarians and humanists, working for the well-being of others, committed to the human as a figure of concern. In their practices, however, the technicality of their procedures makes them apologists for things, even against their intentions.

But this move toward households as a way to ethically classify and explain humans is not new in any way. The household appears in the Western legal imaginary as early as Roman times, when the rights-bearing citizen was in fact the head of the household, not as an individual but in his hierarchical relation to the collective he represented (Arendt 1959). That household was undergirded by kinship, gender, status, and political relations of many sorts, including those of slavery. The household at the center of the human right to water in Costa Rica is different. It is not determined by the number of family members, their gender, genealogy, age distribution, or socioeconomic status. The household of the human right to water is the household of things, an unexpected actor in the political ecologies of rights. This household has the power to quantitatively and temporarily determine the affordability of water in Costa Rica, despite multiple efforts to address affordability through legal arguments about dignity and citizenship.

Neither Laspeyres nor developmentalist economists would likely have anticipated that the CPI would end up working as a nexus between the social struggles for the human right to water and the responsibility of controlling prices in a nostalgic holding on to the principles of the welfare state in Costa Rica. But Sofia and her colleagues afford the consumer price index this new life. By limiting any price increase in water to the inflation rate, regulators affirm the imagined lifestyle of a generic household, marked by its consumption, and insert it into the calculation of a human right. But as we have seen, that household and that consumptive lifestyle is only a statistical purification; it cannot be found in any one Costa Rican household.

With this statistical shift, regulators and international bureaucrats go to the core of liberal philosophy. They first merge human rights and commodities. They determine one through the other when they rely on affordability; a human right is a right to an affordable commodity. And then, after having performed that fusion, they insert a bifurcation, they re-create a distinction by saying that if priced under the 3 percent benchmark, water is indeed a human right and not a commodity—despite being both at the same time. In this context, the price of water (via households) becomes, on the one hand, a contemporary form of humanitarian reason, and, on the other, a way to claim the remnants of mid-twentieth-century welfare ideologies of a state's responsibility to care for its population.

Ana's original demand that we pay attention to the multiple ways in which water is made (un)affordable, while acknowledging it as a gift from Nature and God, has the effect of revealing how the substrate of objects determines the meaning of a human right. Today, humanitarianism depends on floor wax, computer desks, software lessons, and cable TV subscriptions, regardless of their humans. If in Roman times the head of the household represented all of its inhabitants, in the twenty-first century, the collection of purchased things represents all of its humans.

Rivers, rivers where women do laundry, lakes, reservoirs, aquifers, channeled water, ocean water, freshwater, brackish water, water used for irrigation, ice cubes, clouds, waste water. *Items in the water taxonomy produced by Libertarian congressional representatives in Costa Rica between 2002 and 2015*

3 LIST

The previous two chapters have examined the ways in which numerical devices, a formula and an index, help create differences that shape the form of water as a political object, as a material concern of large groups of people. In this chapter I move outside of mathematical worlds to examine another type of device, a list. I also move away from regulatory and statistical agencies to Costa Rica's Congress. At stake is a legal reform to introduce into Costa Rica's constitution wording that explicitly recognizes water as a human right and a public good. Made possible by an entangled congressional procedure, the list that I focus on is a device that turns the recognition of the human right to water into an opportunity to dwell on questions about its materiality. Here, the bifurcation happens in unexpected terms. It hinges upon water's material definition. For congressional opponents of the reform, mainly Libertarian representatives, if water is to be a human right and a public good, its materiality needs to be specified. What substances count as a public good and which do not needs to be taxonomically determined. For supporters of the reform, the specific material form of water is a matter of common sense that does not need to be taxonomically determined. What is interesting here is how, thanks to procedural maneuvers, the discussion of what counts as the materiality of water results in a list produced piecemeal over years. This list inhabits the borders of liberal legal imaginaries as it denaturalizes water as a substance that is subject to property regimes. More broadly, it reveals how ideas about generality/specificity and material stability/instability help clarify what is serious political argument and what is farce. Which ideas count as each is, of course, hotly contested among those involved in the struggle.

To understand the water taxonomy the Libertarians created, I have organized this chapter in two main parts, each with a distinct tone. In the first part, I narrate the procedural life of the constitutional reforms that have been attempted since 2002 to recognize water as a human right in Costa Rica. I also trace the peculiar political place and ethical orientation of Libertarians in Costa Rican politics by providing an abbreviated history of the Libertarian party, examining its origins in 1994, the metamorphoses it has experienced since that time, and ultimately its dissolution. In the second part of the chapter, I return to the constitutional reform procedures to reread them through the words, gestures, tones, and objects the Libertarians have used in their mission to hijack procedure, expand time, and ultimately trouble the material order. The result is a rereading of congressional activity full of ice cubes, clouds, and puddles—a journey through legislative speeches that address new materialist concerns in a manner that is reminiscent of discussions in philosophical salons.

ENLISTING WATER

What is the power of a list? As a display of carefully collected or casually assembled items (humans, countries, songs, foods, species, etc.), how does it affect the world? As precursors of typologies or the culmination of taxonomic desires, lists help organize the world. They punctuate our attention, granting us the possibility of creating simultaneously open and closed groupings. Lists have the capacity to foreclose by delimiting a category, and yet they cannot help but insinuate the possibility of new elements joining, of remaining open. I am interested here in one peculiar list: a water taxonomy. Consisting of thirty-one items, this watery taxonomy revealed the limits of constitutionally designating water a public good. It worked as a device that punctuated the ontological order that congressional representatives and activists rely upon to create a fundamental political separation between a human right and a commodity. It took congressional representatives more than ten years to unwittingly produce this list, a few items at a time. As can be expected, the list's coming to life was wrapped in layers of political spectacle, legal wrangling, activist tactics, and, unexpectedly for me, materialist wonderings. The list was a device that engaged water in a larger battle over what Libertarians frame as state overreach and the threat to the fundamental right to own property.

I began hearing about the items in this list from Eric, an activist friend with a background in philosophy and law who works with one of the oldest environmental law NGOs in Costa Rica. He and his colleagues had spent the previous twelve years stymied by the Libertarian obstruction of their attempts to change Costa Rica's constitution so that water would be recognized as a public good and a human right. Eric's feelings about the Libertarian obstructionism oscillated between shock, revulsion, and anger. Only *diputados* (congressional representatives), he told me, could utter the nonsense the country had been forced to put up with. In Costa Rica, as in many other parts of the world, representatives have a terrible reputation: uninformed and always legislating for their own and their friends' economic benefit. Knowing that if the amendment were put to a vote they would lose, the Libertarian diputados hijacked legislative procedure to make a fundamental point. They were amenable to the idea of a human right to water. But making water a public good and putting it into the constitution represented a "core ideological" issue they were going to oppose at all costs. For them, the state should not own property because "what belongs to everybody, belongs to nobody" and therefore is never cared for appropriately.

This reference to the problem of things that belong to nobody is not original to Costa Rica's political circles. It is a rearticulation of a 1968 article by Garret Hardin titled "The Tragedy of the Commons"—an academic hit published in *Science* magazine to address so-called overpopulation that traveled beyond the academy like few other ideas have. Using the example of pastures and herdsmen, the article makes the argument that common resources (something Hardin defines as resources that are open to all because they have not been clearly bounded as belonging to individual actors) are doomed to be overexploited. This leads to a decrease in the wealth and well-being of those who enjoyed its use without any limitation, but also of society, or the "system" as he calls it, overall. Since its publication Hardin's article has been used to support all sorts of privatization projects across the world. Academics and critical scholars have challenged his argument, noting how the only empirical evidence the article relies upon is an ahistorical understanding of medieval transitions into capitalism (Maurer 1997). Others have documented multiple examples of how common property regimes are structured around explicit and implicit rules that organize resource access and utilization, in many circumstances avoiding depletion and exhaustion of resources (Ostrom 1990; Wutich 2009). In the twenty-first century, the idea of the commons has been revitalized, but the assumptions behind Hardin's arguments are also widely accepted in mainstream legal and economic circles around the world, including Costa Rica's legislative assembly. This revitalization has also resulted in a slippage. Although technically, political scientists, economists, and other social commentators define public goods, common goods, and collective goods differently, in public parlance the terms are used interchangeably.

When Eric spoke of the history of the struggle to recognize water as a human right and a public good, he referred to attempts to bring about *a* constitutional reform, in the singular. But in reality, the activists and representatives promoting the legal change had attempted a series of reforms. Three different amendments had been introduced in the last two decades, each taking up years of legislative procedure. The amendments had the support of many and heterogeneous political groups. Moreover, most everyday people in Costa Rica agree with the idea that in order to secure the human right to water, it must be a public good. I have never heard anybody openly advocate for its privatization. But people do not prefer the notion of a public good because they trust that the state always does an outstanding job managing public goods; they do so because the alternative seems worse. Private appropriation is seen as tied to commodification, profit-making, and *mercantilización* (commercialization), something that, as we saw in chapter 1, people reject in the case of water.

The determination of whether water constitutes a public good is not a new question in Costa Rica's legal system. Water has been legally recognized as public property by constitutional jurisprudence, by the General Comptroller's Office, and, since 1982, by the Mining Code. And yet activists like Eric and many water professionals continue to fight for enshrining this classification in the constitution. They believe doing so would "reinforce," as a lawyer put it, the public property regime under which water exists. The reform would move the protection of water to a foundational level, precluding any future attempts to challenge its public character. This manner of reinforcing the future is a common tactic used by NGOs and activists in Costa Rica. It is one of the most direct ways in which they shape the future even if they do not have a comprehensive vision of it. They believe changing an article in the constitution here and another there can create the preconditions for the yet to come.

But using the constitution to intervene in the future has generated an intense debate about the meaning of farce in public debate and about the role of procedural guarantees in a liberal democracy. Amid these political fireworks emerged the curious list/taxonomy of water that captured my attention. This list, a product of congressional speechmaking and political debate, is difficult to place; at times people found it hard to take seriously, yet impossible to ignore. As a political device with the potential to reinforce the future and allocate collective legal responsibilities, the list has been extremely powerful. And at the same time, the list was only a quasi-event; it never became a full legal taxonomy because it was never adopted, translated into law, or taken up by citizens. In a way, despite being inscribed in congressional records, the list is ephemeral. And yet it was extremely consequential not only for the legal aspirations of the promoters of the reform but also to a more diffuse understanding of water politics among activists and water professionals.

I learned about the list piecemeal, the same way it was constructed. Different people told me about different items at different moments. But after Eric described parts of it in detail, I committed myself to reviewing the congressional records in search of what he and many others saw as a ridiculous and reprehensible maneuver. Reading the speeches given between 2002 and 2013 took me on a rich journey, from depositions given by the foremost constitutional legal scholars and practitioners all the way to raw expressions of embodied disgust leveled by one representative at another. In that journey, however, I found that despite being viewed by many as absurd, the Libertarian list was staged as a "coming together of things that are generally considered parts of different ontological orders (part of nature, part of the self, part of society)" (Thompson 2005: 8). In this coming together, the list became a watery choreography that transgressed the borders that keep things as separate individual entities. The Libertarians put together a wondrous list of types of water that illustrated what they saw as a nonsensical idea: that water could be bounded as public property. To that purpose, the items in their list intentionally combined "technical scientific, gendered, emotional, legal, political and financial aspects" of water to undo any self-evident idea or consensus about what water as a substance is in the first place (Thompson 2005). Overall, their taxonomic list was a curious assemblage that tested its audience's capacity for ontological wonderings, in many cases so aggressively that it shut down the possibility of any engagement between proponents and opponents of the reforms.

Another reason this list caught my attention was its uncanny resemblance to academic concerns of the last decade, which can be glossed under the "new materialist" category. This new materialist literature can be broadly understood as an academic rediscovery of the weight and effect of matter on its supposed own terms, before symbolic meaning.[1] When not reciting their primary talking points—the state's infringement on personal liberties, opposition to the expansion of the state apparatus and to new taxes—the Libertarians crafted an astonishing image of the fluid form of water, a vision of processual nature in constant morphological transition. In a less-than-inspired lyrical moment in Congress, one of the Libertarian representatives told his audience, for example, that *el agua se escapa de los dedos, como el agua en un canasto* (water escapes your fingers, just like water in a basket). But this fluidity was tied to more than human fingers. Libertarians drew on the movement of water across creeks, clouds, the atmosphere, and underground to oppose the possibility of its fixation through the figure of public property. They attuned congressional politics to an unusual heightened awareness of an expanded and fluid sense of materiality. They literally wondered what water is, where its borders lie, and how it can be made discrete so that it can be owned.

In addition to their material wonderings, something else made the Libertarian list and its effects intriguing. The congressional tactics Libertarians used interrupted the linear and forward-moving procedure that characterizes law-making imaginaries. Their list robbed participants in the legislative process of a sense of righteous sequence leading to a future event, to a vote on the proposed amendment. Congressional representatives and observers are aware that the legislative process is flawed and that representative democracy often falls short of its promises, yet the radical disruption of process the Libertarians achieved effected a traumatic break. By interrupting the sequential order of conventional lawmaking procedures, its anticipated temporal and causal linearity, the Libertarians turned a highly ritualized legislative practice into, at best, a quasi-event. They recast it as an occurrence, the historical significance of which was up for grabs, despite seeming to have the direct effect of preventing a vote from occurring in the present. The next section narrates that procedural history.

CONSTITUTIONAL REFORMS

Since 2002, three legislative projects have been proposed to explicitly recognize water as a human right and a public good. Each project had its own peculiar history, and yet they were all under the umbrella of a global humanitarian turn toward human rights as resources for national politics. As we have seen, during the first decade of the twenty-first century, water became a political concern not only for environmentalists, but also for politicians and corporations. And yet, despite that momentum and the considerable international praise Costa Rica's commitments to human rights continue to garner, these constitutional amendments have never been put up for a vote.

Given their symbolic and interpretive weight, constitutional texts are difficult to change. In Latin America, the judicialization of politics that began in the 1980s has resulted in a return to the constitution as a strategic site for the reimagination of life in collectivity.[2] What once were somewhat sacred and inert texts became sets of guiding principles for the elucidation of controversies that everyday people invoke.[3] Not surprisingly, given their impact on legal, economic, and social life, the standards for what counts as a good constitutional text also became the object of broad discussions leading to many partial reforms or complete constitutional overhauls in Latin America at the end of the twentieth century.

Legal scholars and practitioners consider a constitutional text to be of good quality if, among other things, it exhibits the right balance between generality and specificity. A constitutional text has to be general enough to encompass the imaginaries of the present and the future of the political actors involved in its creation—congressional representatives, NGOs, career politicians, civil servants, judges, and some everyday citizens. And, at the same time, it has to be concrete enough so that people can recognize in it enough of a prescriptive program, a set of maxims to guide the fluid "identity" of a collective. As a matter of good legislative technique (*técnica jurídica*), constitutional experts argue for striking the right balance between those two contradictory demands: generality and specificity.

In 2012, Cesar Hines, a public law professor at the University of Costa Rica's (UCR) law school, testified before the congressional committee studying one of the reforms that sought to recognize water as a public good. When a committee member asked Hines to "orient" them about the reform, he responded with an explication of good legislative technique: "The constitutional norm, due to its changing historic context, has to have an open texture, the constitution is a law that cannot be modified every now and then. . . . It must have an open texture so that the legislators can adapt the ordinary laws to the historical context in which they are being developed" (September 24, 2012).

At the same hearing, Manrique Jiménez, another constitutional scholar and also a professor at UCR, stated that, "In producing constitutional norms one has to avoid juridical tautologies, an unnecessary repetition of texts that might lead to confusion. The least room for interpretation a [constitutional] text offers the better it is because, regardless, there are always going to be many opportunities for interpretation and those opportunities need to be narrowed. It is much better to speak of a right, as something that is in the text, consolidated toward the future, than speaking about something like a solidarity principle, which is too loose, too romantic, more sociological than juridical."

Hines and Jiménez posed two competing demands for lawmakers. They asked congressional representatives to not be too romantic or sociological and to produce a text with sufficient precision. They also recommended that the constitutional text should have an open texture, leaving enough interpretive space so that it could adapt to the unexpected circumstances of the future.

Given the interpretive significance of a constitutional text, both in terms of the future and in terms of how it influences the application of any legal norm, it is not surprising that the procedure to modify the contents of the constitution is a complex one. In Costa Rica it requires all sorts of negotiations and alliances to guarantee the support of at least two-thirds of congressional members and to secure the approval of the executive branch. The steps include plenary discussions, committee approvals, voting sessions requiring absolute and compound majorities, as well as multiple "readings'" and discussions of the proposed text across at least two legislative years (May 1–April 30).

The set of procedures that guide constitutional amendments also have to embody important democratic principles. In Costa Rica those principles were translated into a procedural rule that puts no limit on the number of motions or revisions a diputado can request to the text of a proposed constitutional amendment. This opens the door for a type of procedural obstructionism that is often used and that, in one memorable case, led to hundreds of motions to revise being presented to the press in a wheelbarrow.

The first water-related constitutional amendment, which initiated this complicated procedural journey, was introduced in 2002. One of its lead sponsors was Quírico Jimenez, an internationally known dendrologist and conservationist. Most students at the universities where Quírico taught biodiversity conservation knew him well, as did park rangers and conservation managers. He had widely disseminated his knowledge of Costa Rica's trees and their relations within forests. Years of leading workshops, NGO seminars, and publishing scientific papers and popular education documents had spread his name across a broad array of communities. Quírico had never been formally involved in electoral politics. But in 2002 he became a diputado with the Citizens Action Party, the first political group to break the long two-party control over national politics that followed the 1949 constitution. After his work in the Asamblea between 2002 and 2006, Quírico left electoral politics. He returned to his academic and conservation activities, first working on environmental programs for a water utility and ultimately joining the National Institute of Biodiversity, an organization set up in the 1990s to catalogue all of Costa Rica's plant and animal species in preparation for the dream of financing its conservation efforts through bioprospecting contracts with global pharmaceutical companies.

The other congressional member who sponsored the amendment was Joyce Zurcher, a philosophy professor and mainstream politician from the National Liberation Party (the same party whose president abolished the army in 1949, as described in chapter 2, and one of the two parties that alternated control over Costa Rica's electoral politics for nearly fifty years). For years, Joyce taught a required philosophy course for incoming students at the University of Costa Rica. Her subsequent political life also included the positions of national deputy ombudsman [*sic*] and mayor of the third largest city in the country, Alajuela. She belonged to the country's economic elite, and despite being at the center of a number of public controversies, Joyce was one of the first mainstream politicians outside of environmental institutions and NGOs to speak about sustainability, water conservation, and climate change. To date, she continues to be an active member of her political party.

Joyce and Quírico, with the backing of NGOs, activists, and other representatives, decided that focusing on the property designation, and thereby reinforcing water's character as a public good, was a more realistic and effective strategy than simply declaring it a human right. They believed that if water was recognized in the constitution as a public good, activists and the state could resist its commodification and privatization more effectively. Following these political instincts, they worked to include text in the constitution to classify water as a *bien demanial*.

A bien demanial is a legal classification used to denote a good whose existence and utilization must benefit the "common good." In this type of property, what matters is whether a political community believes certain objects or institutions should be tied to "public and collective well-being," as opposed to "private and individual profits." In principle, this means that once an object or legal institution, like a public corporation, has been designated a bien demanial it cannot be traded or transferred via regular commercial transactions. If at some point the state considers that the common good would benefit from privatizing a bien demanial, it can follow a strict procedure to do so. This is what the process of privatization in the heyday of neoliberal reforms in Latin America consisted of. Utilities, television channels, transportation infrastructure, oil reserves, land, and factories were sold to private actors using this kind of legal reclassification. While the authority to lead this process can be deposited in the executive branch, in Costa Rica, only Congress has the authority to move a bien demanial into *el comercio de los hombres* [*sic*] (the realm of commerce of men).

An extensive body of literature in policy studies, political science, and economics, as well as in anthropology, geography, and cognate fields, has examined the configuration of public goods. Scholars have theorized their social lives from alter-globalization, anarchist, communist, community-based, and state-based perspectives (Bakker 2007; Hardin 1968; Olson 1971; Ostrom 1990; Wutich 2009). What is important for my discussion here, however, is that in legal doctrine the character of a public good is derived from its relational activation through sociomaterial patterns of activity, its significance for collective life as evaluated by congressional representatives.

Inherited from a distinction that goes back to Roman law, public goods (*bienes demaniales*) stand in opposition to private goods (*bienes patrimoniales*);[4] the latter are goods that are available for private appropriation and regular commercial transactions. This distinction is one of the first classifications lawyers in Costa Rica learn when they begin their training. The distinction is also one that many non–legally trained people, including congressional representatives, feel comfortable attacking or defending because it fits a vernacular sense of what the state should safeguard for its inhabitants and what it should refrain from intervening in. Joyce and Quírico's purpose in working with that very fundamental distinction between bienes demaniales and bienes patrimoniales was to situate water in the very foundation of the legal imaginary. Making water a constitutional bien demanial was the kind of reform that, they thought, would impact all future legal interpretation. Without necessarily foreseeing its concrete consequences, the way a planner, an economist, or a soothsayer would, this measure would nevertheless inform any upcoming legal and policy measures. It would set an important precondition for the future. The reform would help activists and state officials openly defy any attempt at privatized commodification and would also create opportunities to rectify wrongs. But despite their commitment, they were not naïve. They did not foresee a future devoid of water conflicts in the country, such as already existing struggles over scarcity in Guanacaste, pollution of aquifers in cities, and pesticide contamination of rural community aqueducts (Ballestero 2019). In fact, the need for reform grew out of the rapid multiplication of those water conflicts in both urban and rural areas of Costa Rica. Given their limited power as diputados, all Joyce, Quírico, and their supporters could do was think about the legal conditions they could set up to help address the problems they saw growing but could not precisely solve. To achieve this objective, they focused their reform on Article 121 of Costa Rica's constitution.

The wording of the law and the placement of an article in a particular section of a legal text have important interpretive implications. Diputados in Costa Rica are made aware of this by the nonpartisan congressional staff, who check whether proposed new laws fit with the general principles of *técnica jurídica* (legal technique). Based on those principles, and the training they receive, diputados know that the text physically contiguous to a particular word shapes its legal meaning. This awareness of the implications of the contiguity of words is enshrined in legal interpretation principles such as *noscitur a sociis*, a Latin phrase that has been translated as "a word is known by its fellows" (Tiersma 2005: 121). This principle dictates that when the meaning of a word is unclear, its significance can be determined by looking at the text that surrounds it—its neighboring words. Thus, the contents of an article in a statute or in a constitutional text can be seen as fellowships of words, contiguities that organize exclusions and distinctions in specific ways.

Considering their awareness of the significance of word fellowships in the law, Joyce and Quírico's selection of Article 121 was not inconsequential. The article is a foundational piece of the division of power and checks and balances of the country's democratic architecture. Article 121 translates ideals of sovereignty into concrete functions, jurisdictions, and institutions for Costa Rica's legislative assembly. It states that Congress is responsible for passing and reforming laws, appointing magistrates, approving international agreements, authorizing military troops or vessels to enter the country, suspending individual rights and civil liberties in cases of national emergency, and, among other responsibilities, decreeing the "alienation or application to public use" of the nation's public goods, its bienes demaniales.

As I mentioned before, in Costa Rica the authority to convert a bien demanial into something else, a bien patrimonial, is limited to Congress, although that authority has some gradations. Congress can reclassify most public goods by passing a regular law. There are a few bienes demaniales, however, that not even Congress has the authority to reclassify as private goods. Article 121 of Costa Rica's constitution, the article Joyce and Quírico were targeting for their reform, lists the extremely select group of goods that are beyond congressional reach and can only be privatized by following the complex procedure of a constitutional reform—a process that takes many years to unfold. That group of select goods is buried in the text of Article 121, an article that has grown too long for the taste of legal scholars and judges. There, we find three types of bienes demaniales: (1) all forms of hydropower; (2) coal, sources and deposits of petroleum, along with any hydrocarbons or radioactive minerals; and (3) railways, ports, and airports.

If there is any secular way for the state to establish an inalienable possession, the type of "transcendent treasure to be guarded against all the exigencies that might force its loss" (Weiner 1992: 33), Article 121 is what this looks like. Joyce and Quírico wanted this group of inalienable goods in Article 121 to include water itself. If successful, their reform would prevent water from being moved outside of state control except by a constitutional reform; water would become one of the state's few inalienable possessions.

Once the amendment began its procedural life, its proponents assumed it would sail through the rest of the process. With their optimism, Quírico, Joyce, and all of the NGOs, politicians, and academics that supported them took for granted two things. First, they assumed that "water" was a self-explanatory category. They presumed that anybody could understand, without need of clarification, that stating that water is a public good applied to large bodies of water such as rivers, lakes, and aquifers. They did not think it necessary to dwell on the exact meaning of the word *water*. After all, hydropower and petroleum deposits were also ambiguous categories; their ratios of generality to specificity—openness to precision as Hines and Jiménez explained—could not be easily evaluated. Thinking with Costa Rican legal scholars, we could ask about these categories: Are they too romantic? Too sociological? Closed enough? And yet those kinds of ambiguous categories were already in Article 121. Their second miscalculation was assuming that because many specific laws and jurisprudential sources already recognized water as a public good, the reform would not encounter any significant opposition.

Figure 3.1. Promotional materials produced by NGOs to inform the citizenry about the obstructions that prevented the reform from being voted on.

To everyone's surprise but the Libertarians, once floor discussions started, the reform became increasingly entangled in procedural tactics, intertwined with the nitty-gritty of political ritual week after week. Knowing they were a minority in Congress, the Libertarian party adopted any conceivable procedure to prevent a vote. They did so with such conviction that in 2006, four years after the amendment had been introduced, it was *archivada* (literally, archived). Its congressional life had ended without a vote. With their procedural tactics, the Libertarian party managed to successfully block the amendment, despite the fact that all other parties supported it. Immediately after it was archivada, other issues took over public political discussions. The NGOs that supported the reform redirected their attention to the first referendum in Costa Rica's history, convened in 2007 to determine whether the country would ratify a free trade agreement with the United States and Central America, known as CAFTA (Central American Free Trade Agreement). After a spontaneous political mobilization that has had no parallel in the country's history, the CAFTA referendum was voted on in October of that year. It passed by the very small margin of 3 percent. This reignited fears of water privatization and for-profit extraction by U.S.-based corporations. Many activists stated that one of the objectives of CAFTA was to make water available to multinational corporations that would sell it for profit and deplete the country's reservoirs.

With another Congress in session in 2008, a new attempt to reform the constitution was initiated. This time the lead proponent was José Merino del Río, the founder and leader of Frente Amplio, a self-defined democratic socialist party that, to everyone's surprise, including that of the United States embassy, as revealed by WikiLeaks, garnered third place in the 2014 national election and seated nine representatives, the largest number any leftist party had ever achieved. By 2018, their success had deflated and they only elected one diputado. Their swift ascendance was in great part due to the work that Merino, as everybody in Costa Rica affectionately calls him, had done for years before he passed away.[5] Merino was a politician appreciated by supporters and opponents alike. Born and trained as a lawyer in Spain, he moved to Costa Rica, where he obtained a master's degree in sociology and became involved in national politics. Early on he joined the communist party and was subsequently involved in two political organizations as the leftist parties disintegrated and reintegrated multiple times. Finally, in 1992 he founded the Frente Amplio party. Unafraid to invoke Marx and anarchist thinkers on the floor of Congress, Merino had an adventurous spirit and was a believer in the politics of the street (*política de la calle*), joining in as many protests and demonstrations as he could. Merino was a great orator and worked very closely with NGOs and social movements to channel some of their concerns into congressional discussions. While most people were personally fond of him, in centrist and conservative circles his political ideas were seen as polarizing and radical.

With Joyce and Quírico's project already archived, Merino sponsored a new attempt to include water in the constitutional text in 2008. This time, he considered the legal changes happening all throughout Latin America, where more countries were including the human right recognition in their constitutions. He also took into account the clear signals from the UN that an explicit recognition of the human right to water was coming from the

General Assembly. As a result, the amendment he proposed tied the notion of a public good to the concept of a human right and targeted Article 50 of Costa Rica's constitution, which states that "the state will seek the highest well-being for all the inhabitants of the country by organizing and stimulating production and the most appropriate distribution of wealth."

A relic of the strong welfarist agenda that dominated Costa Rica in 1949 when the constitution was written, this article continues to inspire and support distributive policies while setting off the opposition of believers in individual economic freedom as the highest form of liberty. In 1994, a Congress that successfully passed important environmental reforms added a second paragraph to Article 50, which also states that "every person has the right to a healthy and ecologically equilibrated environment. On this basis, anybody has the legitimacy to report acts that infringe upon that right and claim reparations for the damages caused by its infringement."

This part of Article 50 has been crucial for the legal work NGOs and social movements have done to oppose the operation of transnational extractive industries in Costa Rica, including Canadian- and American-based oil exploration and open-pit mining endeavors. Choosing Article 50 for the water reform was a smart move. It tied water to the "green" imaginary that permeates Costa Rican society and makes Costa Ricans principled, if not always practical, lovers of nature.

But when it came time to discuss the amendment, familiar signs appeared. The Libertarians were again opposing it. They filled the speaker roster and introduced motion after motion to buy time by claiming the need to adjust the proposed text. Again, they took over the floor, creating a sense of circular time, of being back in 2002, as if nothing had changed. Merino and his legislative aides mobilized all sorts of arguments: existing jurisprudence, UN documents speaking about the human right to water, reminders of Congress's moral responsibility to the electorate, the existing consensus among all parties except the Libertarians, the cost of filibustering for public coffers, and so on. None of the arguments, shaming, or political pressure changed the Libertarians' minds. Each time the constitutional reform was discussed, they used their tactics to hijack the discussion and block the vote.

Since then, two more attempts have been made to reform the constitution. But Libertarians still have not allowed any of the reforms in the legislative pipeline to be voted on.

LIBERTARIANS, OF THE COSTA RICAN SORT

The Costa Rican Libertarian party, Movimiento Libertario (ML), was founded in 1994. By 2002 it had become, according to one of its founders and the former chair of the state of Florida's Libertarian Party, "the most successful Libertarian party in the world" (Costales 2002: 40). Thanks to the exhaustion of Costa Rica's two-party system, the deterioration of the public apparatus due to structural economic reforms, and a populist agenda, Libertarians took Costa Rica by surprise. Even though many deemed extreme economic liberalism a minor ideology in traditionally welfarist yet closeted neoliberal Costa Rica, their participation in national elections in 2002 yielded them five out of the fifty-seven congressional seats that make up the Asamblea, up from only one in 1998. This sharp ascent was unprecedented. In the next elections (2006), the Libertarian party would get their highest number of representatives ever, nine. Their sudden success—sudden given the pace of party adherence and conservatism in Costa Rica during the previous fifty years—gave Libertarians deep satisfaction and high hopes of reaching the presidency to transform the country. But in the 2010 and 2014 elections, the ML secured only four seats in Congress. A series of corruption scandals, including the declaration of the party's bankruptcy, eroded the party's gleam as a new option in a worn-out electoral system. But thanks to the procedural "democratic" guarantees the Costa Rican constitution establishes, despite their decline in numbers due to a conviction on fraud charges against the state's election funding mechanism their congressional impact was not completely eroded. The Libertarians continued to exert a major force in Congress until the 2018 election, when too many corruption scandals finally imploded the party and they did not elect any representatives. Many of the economic ideas previously defended by Libertarians are today promoted by a group of fourteen representatives elected in 2018 who identify as the "evangelical bloc."[6]

Following a position they shared with many other political groups of similar ideological persuasion in the Americas, during their political life Costa Rican Libertarians organized to "defend" economic and political liberalism "openly and with pride."[7] When party members explain the history of their organization, they divide it in two. They say they first went through a fundamentalist phase of about ten years (1994–2004) when they made controversial proposals that included abolishing cherished Costa Rican public institutions such as the national electricity institute and the public health system. Sensing the unviability of those ideas, the party became more centrist and entered a second phase (2005–18), one that has included a religious revival, a moderation of Libertarian tenets, and embracing xenophobic discourses.[8] The shift was so clear that despite their name, they stopped identifying as politically Libertarian and began explaining their beliefs as liberal. This movement to the center allowed them to accept financial resources from the Costa Rican electoral system, which they had previously rejected as a matter of principle. This decision infuriated some of the stricter party founders, who ended up resigning.[9] After this diaspora, the organization was left in the hands of Otto Guevara, another of its founders, who has become the most public and prominent figure in the party. He has been elected diputado twice and has run for president five times.

While campaigning for the 2014 presidential election, Guevara referred to his party's ideological shift during a televised debate. He said that he had "left the arrogance of youth behind" and now held "more tempered views." In this "mature" phase, he explained, Libertarians embraced Costa Rica's welfarist history and saw a country that had lost its former well-being due to corruption, excessive regulations, and governmental controls. Guevara continued to "defend private property rights and personal economic liberty," but no longer opposed every form of regulation. He argued for the right amount of it, for a good balance.

But these balanced and tempered electoral talking points of the now-mature Libertarians did not temper their filibustering techniques, nor did they change their perspective on water and its recognition as a public good and a human right. Throughout what Guevara describes as the life cycle of the party, one thing has remained constant: their utter opposition to the constitutional reform to recognize water as a public good. No other legal initiative in Congress has been subject to such a disciplined, systematic, and strategic opposition. In 2015, when the possibility of voting on a law that included the recognition of water as a public good emerged, Guevara told the press that he, by himself, had all that was needed to make sure the law never passed. That blatant inflation of the self, a single man able to stop an initiative that had the support of eight parties, was not unfounded. Guevara's fiery remarks were backed by the procedural loopholes that the Libertarians had learned how to use for years to successfully block any legislation that explicitly mentioned water as a public good.

From a certain point of view, this history seems familiar, particularly for those of us who have witnessed developments in the United States' 2016 presidential election. But that sense of familiarity can be misleading. It makes unremarkable what is not. It obfuscates the possibility of analysis for those of us who feel more comfortable identifying with Joyce, Quírico, and Merino in this story. So in the next section I return to this political history to revisit the sense of analytic disdain that prevents my collaborators from considering the Libertarian tactics worthy of analysis. I enter the world of the list that Libertarians have constructed piecemeal during their speeches opposing the constitutional reform. I will analyze how the list came into being and later focus on its capacity to puncture ontological (b)orders and unleash the scorn of many political actors in Costa Rica and beyond.

LIBERTARIAN ICE CUBES

Costa Rica's Congress allocates three months every year to discuss constitutional amendments. During that period, representatives have one hour, every Wednesday, to address the floor. For those of us who showed up to watch the plenary discussions, the diputados' voices created an uncomfortable situation. We sat in a room designed to provide visual access to the plenary debate but separated from the diputados by a glass window put in place to avoid the multiple interruptions, and occasional insults, that were becoming common when controversial projects, such as free trade agreements, were discussed. The glass prevented sound from moving between *las barras*, the name used for the space where the public sat, and the floor of Congress. To fulfill the democratic obligation of making the discussions available to the public, beyond their live screening on public TV, building managers installed a speaker system that reproduced the crisp and heightened voices of the diputados inside our observation room. But the volume of the speakers was never properly calibrated, turning las barras into a sound box where congressional speeches were so loud, and echoed so much, that they almost made one's body vibrate in unison with the diputados' oratory. Not exactly a pleasant sensation.

Watching the spectacle through the glass was a little dull. I often sat among five to twenty-five activists holding signs and checking calendars to set a date for the next strategy meeting. On one of those Wednesdays, after thanking the president of Congress for letting him address the floor, a Libertarian representative addressed his colleagues with the style characteristic of congressional speechmaking. The first time you hear it, the style is striking. It combines a higher than normal volume, a sense of deep intensity denoting indignation, and a baroque and overly formal etiquette to address fellow representatives: *Señorías, padres y madres de la patria, señores diputados.* . . . In his speech, this representative discussed the morality of water, reciting platitudes and condemning the "suspicious" intentions of the other political parties that supported the constitutional reform, particularly those of Frente Amplio. Those of us who sat in the audience box—activists, everyday citizens, union representatives, and some students—observed patiently. The more debates I saw, the more grateful I was for any unexpected interruption of the tedious pauses, procedural motions, quorum verifications, and contradictory or unrelated speech that characterize the workings of the Asamblea.

On that particular day, after questioning the utility of the notion of a public good and outlining the problems embedded in linking that notion to a human right, a Libertarian diputado fervently asserted, "Compañeros diputados, the text of this constitutional reform is so general that it almost lacks any sense. A reform to make water a public good, stated like that, just as water, encompasses all types and forms of water, it includes the water in these ice cubes, the water in people's refrigerators, and even the water that we get from the cafeteria. We cannot let that happen, señores diputados."

Having lifted for dramatic effect a glass full of melting ice cubes he had on his podium, the Libertarian diputado was painting for his audience a picture of state tentacles reaching into people's refrigerators and workplaces. This fantastical picture resembled a scene from a horror movie more than the sociological images that congressional representatives usually describe when they support or oppose a proposed law. In the observation room, those of us who caught the ice cube statement looked at each other in puzzlement. This clearly was hyperbole. Unsure about how to react, people felt a bit frozen, just like the ice cubes. But soon laughter punctuated the surprise and people came to terms with what they had just heard. A bit of anger started bubbling up. How could a diputado say something like that?

After that Wednesday speech, ice cubes became legendary symbols of the Libertarian legislative maneuvers. They represented the capacity of a single person to stop the procedure and impede a vote. Those promoting the reform began to characterize the whole affair as nothing more than a ridiculous travesty, a deep perversion of the democratic process. The Libertarians were mocking what should be a serious discussion. But underneath the ice cube eccentricities, observers also noted the Libertarians were using these tactics as political maneuvers to protect the economic interests of agribusinesses and other water-based capitalist ventures (e.g., soft drink and bottled water commercialization).

To say that the ice cube affair was memorable would be an understatement. The melting ice cubes in that Wednesday speech were wrapped in two very resistant layers: one is ideological and the other entails clear economic interests. Well aware of this, water experts and activists have borrowed the ice cubes in order to mock the Libertarians. But their scorn has had little effect as the Libertarians are not worried about losing political capital by having their tactics exposed. To the contrary, one can sense how when they are asked about the water issue they take a certain pleasure in explaining their obstructionism. For instance, as she responded to another representative who accused them of hijacking their parliamentary work, a Libertarian congresswoman defiantly said that yes, they had and were going to continue blocking the reform at any cost because they were a "different kind of party, the only ideologically consistent party."

While the ice cube speech was legendary in its own right, ice cubes were just one item in a long list of forms of water put together by speeches that have been used to occupy all the time scheduled to discuss the water constitutional reform during the last decade and a half. By monopolizing discussion time and using procedural rules to make a vote impossible, the small Libertarian caucus has managed to have a major effect on the historicity of politics. For one, they have been able to stop the flow of congressional time, or to put it differently, to expand the present endlessly, increasing its duration to such an extent that they have prevented the future from arriving. In Congress, the future is an event, a moment of closure, the moment of a vote that arrives after the procedural steps designed to keep the democratic ideal running have been taken. If, in this particular case, the present is the discussion of the amendment, the future would be marked by the event of its approval or rejection through a vote. But the Libertarians have undone that sense of legislative consequentialism, by blocking the vote indefinitely, preventing the future of the reform from ever arriving.

We can see how what could, from a certain point of view, be discarded as nothing more than simple and analytically uninteresting obstructionism, from another point of view operates as something of a time machine, a tactic with the capacity to extend the goings-on of the present. It is a wonder-inducing, even if politically troubled, capacity to affect the linearity of time in ways that challenge fundamental rules of collective democratic life. But this capacity to expand time is not solely the result of procedural rules; it

Diputado receta a Guevara botella con agua sucia igual a la que consumen vecinos de Matina

Por **Editor** - 25 Junio, 2014

El contenido de la botella muestra exactamente la apariencia del agua que consumen pobladores de Matina, Limón. Para ellos, es agua contaminada.

Elvis Martínez. El diputado del Frente Amplio, Gerardo Vargas, le recetó a Otto Guevara del Movimiento Libertario (ML) una botella con agua sucia de la que consumen pobladores de Matina, en Limón, en un acto para concienciar al legislador y ex candidato de la Presidencia, cuyo partido se opone a declarar el agua como un Derecho Humano.

Figure 3.2. A Frente Amplio congressional representative gives a Libertarian representative a bottle of dirty water on the floor of Congress.

is not an empty structure. It matters what stories and what words are used to expand the present in this way. It is important to recognize that such temporal expansion is embodied in the words that Libertarians utter in the time allotted for doing so, and those words are temporal wedges. From that point of view, ice cubes, along with a long list of additional items that fill discussion time, acquire a new significance. When assembled into a list, those words are fundamental pieces in the rhetorical and procedural time machine that the Libertarians operate.

AN EVERYDAY LIBERTARIAN

Trying to better understand the reasons why the Libertarians had turned water into such a contentious and symbolic issue, or at least why it had become the substance through which they would assert their ideological points, I ventured into their side of the congressional building. Once there, I met James, a congressional aide in his mid-thirties who has worked with the Libertarian party for several years. When we sat to talk, we did so shortly after 4:00 PM, the time when the official plenary session of Congress begins daily and when the diputado he works for is busy giving speeches or listening to them. We conversed while sitting at a table in the lobby of the party's office. James always made a point of explaining the ideological foundations of what he called the "Libertarian position" on water issues. During our conversations, he also answered text messages, took calls, and continually multitasked. In between those interruptions, his narrative delimited our discursive space to what he called the "ocean" separating the intention of the law from its actual effects. James was a "legal realist." He emphatically asserted that despite everything being said in the press and during political debates, no one had been able to rebut the fundamental argument that Libertarians were making: what the state says it is going to accomplish, it never achieves.

As I heard their "party ideas" I had to concede that it was hard to discuss these issues with James. His capacity to turn the parameters of our conversation into either/or arguments, yes or no statements, was particularly stifling. The state either accomplishes what it promises to do or it doesn't, somebody is right or not, an argument is won or lost. In James's view, nobody had really proved that his idea was wrong, nobody had showed him or the other Libertarian party members that the state is an entity that accomplishes what it promises.

The more James and I talked, however, the more clear it became how particular and contradictory his views of the state were. On the one hand, James's state was a failed entity, an entity capable only of empty promises. But on the other hand, the state was also an entity that remained hungry for more, determined to take whatever it could, regardless of what it had. In this version, the state was a massive and cohesive entity, something like the Old Testament's Leviathan, that snake-like figure that Hobbes borrowed to help us arrive at a sense of unity out of multiplicity and whose tentacle-like extensions could reach, according to James's party colleagues, into a glass to take one's ice cubes away. I was a bit surprised by this. James's invocation of an all-encompassing and clear-purposed state was far from the experiences most people in Costa Rica have in their bureaucratic encounters. More often than not, anecdotes and news of the interactions between people and the state refer to Kafkaesque muddles of contradictory and irrational demands that do not lead anywhere—an experience of bureaucracy that is closer to James's claim that the state is an inefficient entity unable to deliver on its promises.

After a few conversations I came to terms with the contradictory figure that James and his colleagues see. I needed to make sense of the Libertarian opposition to the joint recognition of water as a public good and a human right from within, rather than against, their contradictory philosophy of the state. I needed to learn to think through a vision of expansive shortcomings, a state that is failed and inefficient as well as ambitious, ravenous, even monstrous. This was a philosophy that not only dealt easily with, but thrived on, inconsistencies. If I had limited myself to diagnosing contradictions and a lack of discursive cohesion in their views, I would not reveal much about the Libertarian tactics. Focusing on contradiction and lack of cohesiveness would reveal more about my demand for consistency than anything else.

This was made clear on another occasion, when James was explaining his view of the state and its expansive shortcomings, and he brought to my attention another register that is crucial to understanding his party's ideas. He was emphatic in explaining how the only "realistic" position on the topic of water was theirs. All he and his party had done, he said, was to take an "empiricist" approach to the question of public goods. If the Costa Rican bureaucracy has proved unable to secure rights and obligations currently on the books, why do proponents of the reform expect things to change? How can the state enact the figure of public property beyond requiring permits and paperwork? In our back and forth, both James and I got exasperated. After I apologized for insisting for the third time on the question of realism and why he thought the failures of the state would inevitably be repeated in the future, he responded by telling me not to worry, he understood that I was "playing the role of *abogada del diablo*" (devil's advocate) and that it was fine, "but," he continued, "look around you, this is how things are and will continue to be. These are the facts." No why or when questions were necessary to historicize the world James witnessed. For him, things were just the way they were, and plainly so.

This was the first time during my fieldwork among all sort of experts and activists (economists, hydrologists, lawyers, sociologists) that I had to navigate a discursive world that admitted no change whatsoever, a vision of what is so tightly assembled that it left no space for the possibility of alternative interpretations. Our conversation left me with a sense of how peculiar my anthropological take on the world as susceptible to being "opened up" to "new" ways of thinking was. Unable, or unwilling, to find space for that view, James saw no need for opening things up or understanding them anew; things were as they were, and all I needed to do was accept that fact and become an empiricist, just as they were.

This view of history, and of the density of what is, turns Libertarians into what I think of as literalist ontologists, people with beliefs about collective life that rely on a literalist display of what they conceive as self-evident fact.[10] This literalism depends on three assumptions: First, it presumes there is little or no ambiguity or room for interpretation in the meaning of facts. Success is success, failure is failure. Second, it sees in the world rich factual evidence that is available for apprehension if people pay close attention. For that reason, digging deep for causes is unnecessary; facts lie explicitly around us. And third, to act and think politically is to place facts in the categories to which they naturally belong—a classificatory relationship that is straightforward and that can handle contradictions very comfortably.

With a sense of this literalism in place, and some background on the politics of constitutionality and public goods in Costa Rica, I now focus explicitly on the contents of the Libertarian water taxonomy and the list that embodies it. As I mentioned before, if procedural rules are the conditions of possibility for the Libertarian manipulation of temporality, the words that associate item and category, of which ice cubes are the item and water the category, are the possibilities embedded in those conditions. In other words, the Libertarian semiotic dance has a fundamental component that is not automatically granted by the expansion of time they accomplish through congressional procedure. Thus, it is not enough to merely map how they expand the present by avoiding the future. The semiotic component that makes their maneuvers extraordinary resides in the actual words they utter, the substance of that newly created time, the words through which they make up their taxonomic list. As we will see, that list does not lead to a simplistic sense of water. On the contrary, despite its farcical tone, the list reminds the audience of the fluidity of form and the constant change that water embodies. This reminder is traumatic because it disintegrates the public property regime as a logical possibility, robbing NGOs and supporters of the constitutional reform of the legal instruments on which they depend for their democratic imaginaries.

WATER TAXONOMIES

As I mentioned earlier, congressional floor discussions are for the most part dull happenings. If observers at las barras found the weekly constitutional discussions tedious, one could only imagine what it would be like to spend a good amount of your working hours there every day. After many hours of watching their formal proceedings, I was not surprised to see the means diputados found to entertain themselves. The public used to learn about these tactics from newspapers that published images of representatives sitting in their *curules* (seats) reading the newspaper or, in a few egregious cases, asleep with their heads tilted backward or fallen forward, lax, and deep into the netherworld of dreams, succumbing to the lullaby of turbulent speechmaking in the background. These days, diputados have cell phones and laptops to make more efficient use of their time while their colleagues address the serious issues facing society. Of course, when an attack is made against a party or a law that is of special interest, things are much more lively.

The Wednesday sessions when diputados were required to discuss constitutional reforms became, over the years, affectively charged meetings. The procedural drag that had stopped the constitutional reform from moving in any direction created a deep sense of frustration among other congressional representatives and water activists. For many diputados, the situation was difficult to make sense of. But even worse than that was the individual pain of listening to the Libertarian speeches on Wednesdays. If initially people were startled by their words, they soon became old, dreadful, even torturous. Whenever other diputados were lucky enough to get the right to speak, they tried any strategy they could to exhort the Libertarians to stop talking, to be quiet, to stop wasting everybody's time. Mauren Ballestero was a congressional representative from the same party that Joyce Zurcher belonged to. While not related to me in any way, most water-related people I met very quickly asked me if we were related, as if to clarify my political and moral allegiances. Mauren is an internationally renowned water "expert" and a strong supporter of a new water law that was also in the works. She publicly confessed to her colleagues during one of the sessions that on Wednesdays she came to work reluctantly. She was tired of hearing the Libertarians' "imprecise arguments" and outrageous misrepresentations of reality. She was tired of being forced to hear *llover sobre mojado* (rain over water), to witness Libertarians "take a topic, delay it, and discuss it as if it were a real topic." She was sick of seeing reality being mocked.

It was not hard to see where these feelings came from. Putting together their water taxonomy required hours and hours of Libertarian oratory challenging taken-for-granted, foundational legal and material ideas. Take the following examples of how the Libertarians slowly produced their water taxonomy through years of congressional debates. Mario Quiros, the author of the ice cube speech, told his colleagues on another occasion, "I remind you, señores diputados, if we want this [constitutional reform] to mean anything, speaking of water as a general category that includes all waters—the water in the ice cubes, the water that flows through pipes, waste water, and even sludge—will not be helpful. Speaking in these terms would mean that any and all water would always be state property."

On another Wednesday, Evita Arguedas addressed the irrationalities of declaring water a public good and augmented the Libertarian taxonomy. Referring to issues of generality and specificity, she declared, "as we see, [the proposed project] does not state what type of water we are going to protect, and here is where the issue becomes dangerous, ladies and gentlemen, will it be ocean water? Maybe fresh water? Brackish water? The water used for agriculture, animal husbandry, or the rivers where women do their laundry? The reform we are discussing does not specify that anywhere."

Another day, the Libertarian holding the right to speak focused on the intense consequences of individual categories, rather than putting them side by side. He chose to deepen the logic behind one taxon, water in the form of clouds, and proceeded to put forth a hypothetical case: "If water was a public good, an airplane that travels through the air couldn't do so because it would pollute the clouds and it would need a permit to do so. . . . This is the absurdity of this project."

But the most outrageous item in the Libertarian water taxonomy was yet another one. For Libertarians, along with ice cubes, ocean water, clouds, water in canals, and so on, another body ran the risk of expropriation by the state: "Fellow diputados, you and I are more than 70 percent water. Our bodies are more water than anything else. If this reform were to pass, you and I would be mostly state property!"

Throughout the years, speech after speech, Wednesday after Wednesday, these kinds of statements yielded a taxonomy of thirty-one types of water. Thirty-one instances that revealed the impossibility of taking water as a general category upon which to impress the designation of a public good. The more you knew about the speeches, the quicker you were to dismiss them as nonsense, as ideas belonging to what has recently been named the universe of "alternative facts" after political developments in the United States. And yet the list does much more than merely exemplify the "irrational."

For instance, we can think of the resemblance between the Libertarian taxonomy and a Chinese encyclopedia created by Borges and popularized in Anglo-American academia by Michel Foucault.[11] Foucault uses Borges's encyclopedia to make a point of the limits that unfamiliar, even unintelligible, classificatory systems so helpfully reveal. The encyclopedia Borges describes, titled the *Celestial Emporium of Benevolent Knowledge*, offers a taxonomy of fourteen categories of animals that include: those that belong to the emperor, embalmed ones, those that are trained, suckling pigs, sirens, fabulous stray dogs, those included in the present classification, and so on. Foucault finds the "wonderment" in Borges's classification in that it creates a symptom, a hook that we can grab to analyze "the limitation of our own [system of thought and] the stark impossibility of thinking *that* [limitation]" (Foucault 1973: xv; emphasis in original). For Foucault the list makes us uneasy not because it puts unexpected and maybe noncredible entities next to each other. Rather, the encyclopedia demands an account of the grounds on which its classification stands. In a similar way, the list makes us uncomfortable because it robs us of the coherence of knowing its foundations implicitly; and at the same time, the sense of incoherence and uncertainty it generates saves us from having to think seriously about its alternative foundations, allowing us to bracket its seemingly irrational grounds.

The Libertarian taxonomy asks questions about the grounds on which fellowships of words can be put together, questions about the taken-for-granted foundations of a taxonomy of public goods. How can notions of public property, human rights, and water itself capture the typological richness of water? The Libertarian list questions those foundations via the material familiarity of water, its embodied dimension, asking what kind of substance could water as a public good be exactly. In other words, besides effecting a procedural interruption that extends time, the Libertarians' list reveals how ambiguous the water category is as an object. The list reminds us of the material properties of water, invoking its fluidity, its obsession with form-changing, disabusing us from the habit of bracketing such fluidity. Because of this material-semiotic reminder of its indocility, the list is taken as nonsensical, as an extravagant interruption of a political and legal process that, in order to work, needs to suspend any literalism about the materiality of water. If the legislative procedure is to work, the material category of water needs to remain vague. For that reason, and given that it embodies a proscribed literalism, the list becomes a farcical maneuver, a cynical move that supporters of the reform can only encounter with incredulity. For the proponents of the reform, and opponents of the Libertarians, the literal material richness of water belongs elsewhere, certainly not in Congress.

LIST: A WATERY POETICS

> Ríos, ríos en los que la mujeres lavan la ropa, lagos, represas, acuíferos, agua canalizada, agua de mar, agua dulce, agua salobre, agua usada para irrigación, cubitos de hielo, nubes, aguas residuales, aguas negras, agua embotellada, lluvia, 70 porcentaje del cuerpo humano, agua consumida por animales, agua llovida, agua entubada, agua en refrigeradoras, agua en cafeterías, charcos de agua, agua subterránea, agua usada para lavar ropa, agua para tomar, vasijas y otros objetos que contienen agua, agua para bañarse, agua para bañar los caballos, vapor.
>
> Rivers, rivers where women do laundry, lakes, reservoirs, aquifers, channeled water, ocean water, freshwater, brackish water, water used for irrigation, ice cubes, clouds, waste water, sludge, bottled water, rain, 70 percent of the human body, water consumed by animals, piped water, water in refrigerators, water in cafeterias, water puddles, rainwater, underground water, water used to wash clothes, drinking water, vessels and other objects holding water, water to take baths, water to wash horses and cattle, vapor.
>
> Libertarian water taxonomy developed through more than ten years of political discussion in Costa Rica's Congress and compiled by the author

This list was assembled and used in parts—subgroups put together and mobilized as congressional speech and as evidence of the Libertarian immorality. Ultimately, the full taxonomy consisted of thirty types of water (see above). Types such as drinking water and water in pipes, for example, might be taken as redundant, and yet, poetically, the difference between them is enough to warrant their individual presence in the taxonomy. It is to that poetic richness that I want to direct our attention now. What kinds of statements brought this fantastic list to life? And how can we make sense of what it accomplishes?

Lists are prime spaces for the negotiation of the contents of categories.[12] Through their notation, lists communicate struggles over their particular scope. Their itemizations function as semiotic maps whose contents and borders confound the list's seeming precision. Itemization gives lists a contradictory and powerful capacity to be open and closed at once. As we consider the specificity of a class—given by the elements being listed—we inevitably question whether there are other elements that should be included. Are there too many or too few items? What are the twists that led some elements to fit in and others to be excluded? Despite any aspiration to be an affirmation, a list is always a question: thoughts of whether elements are missing, have been excluded without reason, or have been included without deserving to be, are intrinsic to our understanding of lists.

This simultaneously open and closed nature grants lists the power to announce multiplicity without having to narrate it explicitly. One way to make sense of this power is by following Roman Jakobson's thinking about the poetic function of language. For Jakobson, one of the functions of language is poetic, a form of meaning-making that does not follow the usual patterns and rules that in its denotative function we abide by. When organizing language to privilege its denotative function, Jakobson tells us, we combine two axes of sense-making: *selection* and *combination*. Selection is guided by principles of similarity and dissimilarity, the meaning that makes signs synonyms or antonyms of each other. This principle allows us to produce statements that literally "make sense" by avoiding contradictory or paradoxical statements when the purpose is to denote a particular content. At the same time, such selections are organized through a second axis that orders the combination of signs guiding their spatial and temporal contiguity, that is, by putting next to each other words that follow a certain sequence that makes them intelligible. For Jakobson, the poetic function of language projects one axis to the other, moving the emphasis of meaning-making "from the axis of selection to the axis of combination" (Jakobson 1960: 358). That is, in a statement where denotation is the main function, a sign's dominant purpose is to facilitate the relation of similarity/difference in meaning (its selection axis). If, instead of denotation, the poetic function is dominant, the emphasis of meaning comes from elsewhere. In the poetic register, meaning is largely determined by the surrounding companions of a word (its combination axis), by their spatial presence and place in a fellowship of words, rather than by the relations of similarity/difference between each sign. That is why we can talk about things like a flying car, fluid stone, or obscure sun. In this sense, seemingly contradictory words, signs that literally would make no sense when put next to each other, acquire their significance and more than that. They are essential for poetic meaning-making. To put it differently, in the poetic function of language, meaning is commanded by the place of a word in a fellowship of words, even if the principle of selection tells us that the meaning of such neighboring words is contradictory or nonsensical. In the poetic function, the combinatorial logic provides enough sense to make the statement meaningful. These different emphases in the meaning-making process, poetic and denotational, are pushed to interesting limits when we think about lists. In a list, what matters is the placement of an item within a category, as a token of something larger, even if that placement seems contradictory. We could describe this particular distribution of the emphasis of meaning-making, from selection to combination, as a form of "technical poetics" (Paul Kockelman, personal communication, 2016).[13]

Seen in this way, "a list is thus a system of relations between elements, the elements being both the contents of any list . . . and the . . . urges to which they are linked" (Philips 2012: 99). While many lists are conceived vertically, as denotational devices that emphasize hierarchal relations between a category and its tokens, other lists are conceived horizontally, arranging "incommensurable elements that lightly touch, snap into an assemblage, or simply differently comprise the texture, density, tempo" of people's world-making practices (Stewart 2013: 42). The water list that the Libertarians produced combines both characteristics at once. It is a vertical effort to specify the meaning of a category—water—in order to trouble its potential legal interpretation. And yet it is also a horizontal list that only momentarily clasps its elements to each other, a fellowship of words that challenges common sense by, for example, putting in the same category underground water, water used to wash clothes, sludge, and water in cafeterias, and responding to the texture of the Libertarian's world- and time-making practices.

On the basis of that capacity to puncture a taken-for-granted denotational function, the Libertarian list returns our attention to the semiotic difficulties of creating a separation between a commodity and a human right by making a general category, such as water, precise. Such difficulties are often ignored; people can go about their daily lives without having to analyze them. But this list creates space to wonder about those difficulties. The list queries the meaning of water itself, quietly producing a technical poetics of the ontological order to which water belongs. Here, the attempted bifurcation that proponents of the reform want to create, the separation of a human right from its appropriation as a private good, seems out of place. From the Libertarian point of view, it is improvident to classify clouds, 70 percent of the human body, and sludge as public or private property.

To many observers, the Libertarian list and temporal tactics are nothing more than a curtain to obscure their long-standing economic interests. For that reason, their political adversaries assign little to no value to the Libertarians' doings. Critics reduce the Libertarian list to nothing more than a tactical maneuver in a process of political negotiation that has been coopted by mediocrity and vested economic interests—a process everybody, participants and critics alike, seems to already understand thoroughly. The list embodies all of that, but not only (de la Cadena 2015). It is a tactical instrument to protect particular class and economic interests, but it is also a device that reveals our material ontologies and their limits. Those limits become visible thanks to the taxonomic ordering that the list effects. The list produces a "shock of illumination" (Raffles 2002: 42) that consists of more than revealing how a taxonomy is flawed, how the items it consists of fail to capture what people see in in the world. The shock of illumination comes from making visible the assumptions on which a taxonomy is built. It directs our attention to the poetic coordinates that undergird what we take for granted as self-evident, and for that reason in need of no explanation, much less discussion. This shock reminds us that "the value of classificatory categories lies in more than their dismantling" (Raffles 2002: 42).

CONCLUSION

By performing a type of illumination that can be achieved without dismantling, I have wondered about the politics of water as a way to make sense of a world where familiar categories are constantly, albeit temporarily, upended. The Libertarian list is a device that requires one to come closer, inspect what seems repulsive, and elucidate how difference emerges out of its dark corners. It allows us to wonder about what is at stake in procedure and how time can be expanded or contracted, but it also allows us to notice how the words uttered in a controversial process have material implications. As

I have noted, for many the Libertarian list deserves only rejection. And yet that frustration and antipathy for something they take as "irrational" are symptoms of where the limits of reasonability, of the taken-for-granted, are placed. I take this list as a poetic oddity in and of itself, a device that punctures inherited categories, separating substances in unexpected ways, if only speculatively and for dramatic rhetorical effect.

Throughout this chapter I have put forth two main points. First, I have shown that to the extent that there has been a controversy over the human right to water in Costa Rica, it has been around the question of property regimes, the economic and political implications of the notion of a public good. Second, I have shown how the list that emerged from the Libertarian filibustering tactics was a device that revealed the politics of the material borders of water, ultimately also showing their material impossibility. The taxonomy revealed how, to the extent that ideas such as public goods continue to matter for water, they depend on the fiction of an orderly partition of water bodies.

By extending their literalist aesthetics to the materiality of water, the Libertarians challenge the ambiguity of that legal fiction and they erase a generality that is necessary to establish the bifurcation between a public and a private good in the first place. The Libertarian poetic materialism provides an excess of specificity that neither legal technique nor political discourse can handle. That excess turns the list into an oddity, a strange formation that offers too much texture and, for that reason, is out of place in constitutional debates. This is one of the reasons why supporters of the constitutional reform deem the list irrational, farcical, and—to put it bluntly—stupid, nothing more than a smoke screen to hide familiar economic interests. The significance of that struggle is broad; it goes well beyond water. The task of defining a concrete poetic ratio of generality to specificity—openness to precision—underlies many fundamental discussions about the limits of the possible, in Costa Rica and elsewhere.

In their book *Thinking with Water*, Cecilia Chen and colleagues (2013) propose attending to how "[w]aters literally flow between and within bodies, across space and through time, in a planetary circulation system that challenges pretensions to discrete individuality" (12). They invite us to engage the "aqueous, [by] actively questioning [its] habitual instrumentalizations" (3–4). In a disorienting turn of events, the Libertarian list does just that. It compiles material states and water uses that escape water's habitual instrumentalizations. The list is a fellowship of words that transgresses any pretensions to discrete individuality as a way to prevent the subjection of water to legal categorizations stuck in what Libertarians strategically deem an old scheme of the partible—the distinction between public and private property. As one of the Libertarian representatives told his colleagues, the fluidity of water makes the concept of public property . . . nonsense. This might be because we are using the wrong concept to regulate this issue, the wrong legal instrument; we are using the concept of public or private property, an invalid, inapplicable, and inconvenient dichotomy to regulate water resources—I would even say an obsolete dichotomy.

The ways Libertarians challenge ideas of public goods and human rights remind us that reimagining materialities is not intrinsically a more democratic, decolonial, non-neoliberal, nonpatriarchal world-making project. The Libertarian list might even reveal that some of the excitement over the remaking of the material world in many academic grounds might be tied to a particular moral and political optimism, which is only possible from a certain position; that optimism is far from being generalizable as an ontological condition. Encountering Libertarians as new materialists and transgressive ontologists reminds us that probing an ontological order is always risky. Following their tactics also makes visible how elastic that order is, how far it can be stretched until it snaps, taking with it many of the technolegal devices we use to try for more democratic futures. If the current liberal order snaps, it takes away the separations that help tweak the world-that-is toward the world-that-we-hope-to-see. In this sense, the Libertarian list is extremely generative. It effectively reveals the limits of what is thinkable in the legal and political setting where the distinction between a human right and a commodity is mobilized. And yet it also reminds us that undoing that distinction is never a clean process. It is an inevitably messy one.

I dwell on this odd list, its authors, and its temporal implications in order to understand how people venture to challenge taken-for-granted orders, even if it is only rhetorically and as a means to protect specific economic and political interests. I believe such moments reminds us that separations, and not entanglements, can also be democratic figures in particular times and spaces. In this case, we see how unassuming devices, such as lists, allow people to push for or against such separations in the process of challenging or resisting new meanings to water as a political and legal category of material-semiotic sense making. After all, the capacity to organize the world taxonomically through lists continues to undergird constitutional moments where the legal order, and with it the worlds within which it exists, are being remade. Lists matter greatly in the world and thus examining their making is a way to grasp what it takes to inhabit the world as it is, but differently. Lists are good to think with, but also good to act with.

The doubts Libertarians cast on public property as a legal instrument give pause to the self-evidence of water by defamiliarizing what is usually bracketed in constitutional discussions, its materiality. Water has proven historically difficult for the clean adoption of property regimes (Bakker 2007; Bluemel 2004; Boelens and Zwarteveen 2005; Radin 1996). What is of interest in this case is how the recognition of the material unwieldiness of water makes explicit how legal procedures are not equipped to grasp the challenges its unruly materiality poses. The Libertarian list attends to the physical continuity and constant challenge to fixed form that water effects. The categories of commons, public or private goods, or the classic distinction between *bienes demaniales* and *bienes patrimoniales* imported from Roman law are not enough for ice cubes that change physical state, clouds that move, or 70 percent of human bodies. This peculiar continuity of water as a substance is turned into a tool to destabilize the metaphorical and legal bounding that objects of property are predicated upon. By reminding us of the material fluidity of water, the legal fiction of its transformation into a clear object of public property is questioned by Libertarians and affirmed by supporters of the reform.

So far we have looked at three devices, formula, index, and list. While we can find these devices across most areas of concern for collective life across the world, I have examined their lives in Costa Rica, a country where, for at least two decades, state norms and institutions of collective life have been taken as static, not completely immutable but difficult to change. For that reason, those promoting the recognition of the human right to water have opted for tactical shifts that are not large scale. Instead of challenging the very foundations of legal and economic institutions, they preserve them as a way to maintain some of the welfarist legacy and liberal property regimes that characterize the country. This reticence about all-encompassing change has slowed down structural adjustments and austerity measures and, in the process, has allowed the country time to wonder about whether self-transformation is something that can be collectively accomplished.

The next chapter shifts our attention to a different place and kind of device. I take us to the state of Ceará, in northeast Brazil, where water activists, technocrats, and politicians have large-scale ambitions. In their efforts to secure more democratic water access, they visualize an all-encompassing shift. They organize to change society as a whole by promoting a transversal responsibility to care for water. They work with another kind of device, one designed to unleash comprehensive transformations, despite the impossibility of verifying all of its effects. That device is a pact, a legally nonbinding aggregation of promises to care for water that, while consisting of moral and intimate commitments, differs significantly from intimate forms of responsibility for kin. I now turn to examine that device to explore its capacity to create collectives without relying on notions of belonging or membership. As I will show, in Ceará, the future of water takes form through a loose collective, via a unique type of gathering.

4 PACT

At the beginning of a Water Pact (WP) meeting held in a community center in rural Ceará, in northeastern Brazil, the lights were dimmed and a promotional video was projected showing striking aerial views of reservoirs, agricultural fields, close-ups of children playing with water, animals drinking from ponds, and irrigation hoses dripping water next to seedlings beginning their lives in the *sertão* (hinterlands). Commissioned by Ceará's legislature, the video captured the state's dramatic environmental and economic conditions. The semiarid environment where Ceará is located is a land of extremes. It has the capacity to reinvent itself into a lush landscape after a few rainfalls, only to return to its brown, orange, and gray palette once the seasonal rains pass. The images in the video portrayed these extremes beautifully. The narrator, a male voice of the kind one associates with radio announcers, spoke about the state's efforts to bring water to all of Ceará's citizens. He emphasized that all of the projects and initiatives designed to make water accessible grow out of a single conviction: water is more than a need, it is a human right.

I was attending this meeting as part of my fieldwork in Ceará, where I had come across the Water Pact, a new effort to deal with water scarcity and realize the promise of securing water access for all. With a variety of legal and institutional reforms in its history, the state of Ceará had decided to launch the pact as an unprecedented initiative that would make water a "transversal" concern of all citizens. Originating from the aspiration of a group of *técnicos*[1] (technical personnel), the pact was designed to disrupt a historical approach to water-related events—floods and droughts—characterized by relations of patronage (Ansell 2014), exchange of political favors (Lemos and Farias de Oliveira 2004), and politicians' search for the personal glory of finally solving the state's perennial water problems through spectacular infrastructure developments. I quickly became fascinated with the impetus of the pact and its aspiration to make water a transversal concern, something everybody felt responsible for. The pact's objective was not to adopt a legal definition or change a formula to secure the humanitarianization of water, as we saw in previous chapters. Instead, the aim was to change "society," to remind people of their capacity, and obligation, to *care* for water, a substance that in Ceará has the power to redefine the politics of justice, nature, and life itself.

The meeting where the promotional video was shown was one of dozens of meetings organized to create the pact. According to official records, by 2009, two years after the original idea was conceived, about eight thousand people and eighty-six institutions had been involved (Assembléia Legislativa do Estado do Ceará, and Conselho de Altos Estudos e Assuntos Estratégicos 2009: 17). At the end of the process, in 2010, the organizers celebrated that more than 136 municipalities, twelve watersheds, and the state of Ceará as a whole all had their own particular pacts. Practically speaking, the pact consisted of a series of public meetings organized as promise-making rituals. At those meetings, participants committed to act ethically and do what they could to deal with water issues in more responsible ways. They promised to care for water. If historically a lack of water stopped Ceará's residents from moving along the line of "progress and sustainability," in the future, care for water would propel them into more just and sustainable living conditions.

The pact was predicated upon a simple idea: a promise to care that reflects the specific conditions of people's lives is a promise that people keep. Unlike the law, which people in Ceará believe creates obligations beyond one's specific conditions, the pact would not replicate a central or universal authority. It would emerge from meetings where people would publicly commit their time, resources, and affects. Those pledges would be written on slips of colored paper, displayed on whiteboards and walls, and then transformed into electronic documents. The only universal thing about the promises was their material form, their notation on a slip of colored paper. Otherwise, each promise was unique, attached to the context from which it emerged. That specificity, the pact organizers believed, would change the future historical narrative about water, loosening the plot line most people use to explain the state's water history: a harsh semiarid environment, a legacy of dispossession, and ultimately Ceará's people as *um povo sofrido* (a people that experiences sustained suffering).

As it turned out, despite the organizers' serious attempts to reach a broad cross-section of society by gathering promises to care for water, most promise-making rituals reached a very narrow group of people. Mayors, municipal appointees, local politicians, state secretaries, regional technical personnel of various federal bodies, congressional aides, and a few regional directors of NGOs filled the participation rosters. A pact that was designed to change society turned out to be a mechanism to bring the state closer to the state itself; it became a tool to rejoin the task of governance with one of its own conditions of possibility, the particular individuals willing to take on the task of making the state responsive.

How would an anthropological analysis of the strange formation that is the pact look? It is tempting to begin from a hermeneutics of suspicion, an approach that would document the ways in which the objectives the pact organizers set for themselves are betrayed, wittingly and unwittingly, by the complexities of collective histories of dispossession, the pressures of late capitalism, or the drama of climate change. I will take a different approach, however. My collaborators are fully cognizant of the risks associated with this kind of project. They know how recalcitrant the history they are trying to change is. They are aware that, in a sense, they are setting an impossible task for themselves, and yet, they persist. They still attempt to unleash a series of acts to reshape water access and attain the humanitarian goal of universality. So instead of documenting their failures, I am interested in how they imagine the pact can transform the world of water.

To trace the work involved in creating the pact, this chapter addresses three main issues. First, it places in historical context the uniqueness of the pact by first describing the history that its organizers want to break with, and then referring to the more immediate origin of the idea of creating a pact. Second, I take you to different moments in the planning and implementation of the pact to show how its organizers' insistence on affirming multiplicity and specificity, and the special attention they give to contradiction, makes the pact depart from more familiar methods of organizing collective life, such as resorting to new legal orders. Finally, the chapter examines what is at stake in the form of the pact. I examine its material form, its original inscription on colored slips of paper that remind one of Post-its, and its conceptual form, as an aggregation of specific promises that remain tied to the context in which they were made. By taking this form, the pact is expected to effect two separations or breaks: one between the past and the future and the other between the political commodification of water and its becoming a humanitarian right accessible to all. Overall, the pact proposes a form of activating collective care that does not presume belonging and that is practically impossible to verify. I end by reflecting on this impossibility and by considering how to engage with historical transformations when we do not have the luxury of empirically verifying their consequences.

A HISTORY TO BREAK WITH

In Ceará, the prominence of water as a political substance is the result of environmental, economic, and political conditions. Located in a semiarid environment and subject to deep temporal and geographic variabilities in rain patterns, Ceará constantly experiences water deficits—needing more water than it annually accumulates through rainfall. This degree of scarcity has turned water into a valuable economic and social good that has historically shaped political and economic structures.

In the nineteenth century, one of Brazil's first republican governments built the Açude do Cedro (Cedro Reservoir) in Ceará, in response to the 1877–79 drought—an episode that killed more than 500,000 people (Greenfield 1992; Lemos et al. 2002). The Açude do Cedro has been since mobilized as a symbol of the state's commitment to help the population deal with lack of water. The structural reasons that explained people's vulnerability to drought, while acknowledged, have remained in the background, and since then infrastructure construction has become the primary "solution" to water problems.

Despite how much variation there is in how people in Ceará experience water lack, there is a shared story that almost everybody articulates about its origins. The cyclical recurrence of water crises—it is estimated that six in every ten years are water deficient in the northeast—creates opportunities for redeploying that shared history. Sometimes people add a new chapter, but even if they do they return to the figure of the *coronel* and to the political practices of clientelism as explanatory devices. The Water Pact was imagined as an opportunity to break with that history. Many people explained that, ultimately, what they were trying to change with the pact was the legacy *coronelismo* left in Ceará.

Coronelismo is a form of political organization that depends on negotiations and compromises between state governments and *coroneis*, representatives of local elites who were conferred that military rank, or in some instances spontaneously assumed it, when they became responsible for large landholdings during Brazil's imperial history (Kottak, Costa, and Prado 1996). The power of the coronel was based on his violent coercive methods, his leading bands of armed men, and his economic assets. Throughout Brazil's history the coronel mediated between the peasantry and municipal and state authorities, functioning as a bridge between people's needs and the public resources funneled into the region from the state and federal governments (Tendler 1997). Coroneis remained effective political figures during republican life and continued to bargain with the state and federal governments to personally benefit from public programs and infrastructure construction (Barbalho 2007; Carvalho 1997). In the 1930s coroneis were disarmed by the federal government, but their social role did not disappear.

After their disarmament, coroneis retained their prominence through a series of clientelist practices they used to filter economic benefits directed to the disenfranchised majority (Chilcote 1990; Furtado 1998). Mayors adjusted public programs to create alliances with coroneis, who in turn used their influence with voters—especially the wage laborers and sharecroppers working for them—to ensure enough votes to keep preferred mayors in office. The periodic transfer of resources to fight drought deepened these clientelist practices. Jobs, with lower than legal wages in some cases, and emergency supplies were assigned to the politically loyal and reservoirs, dams, and channels were built on private properties where they were inaccessible to large parts of the public unless they became allies with coroneis and other local bosses. This system, known as the "drought industry," made people in Ceará aware that water works and emergency aid created subject positions, with specific benefits and redistributed relations of debt.

At the end of the twentieth century, that configuration of power, public works, votes, and peasantry began to change. International financial institutions, progressive politicians, multinational corporations, new research institutions, and innovative federal policy frameworks, such as Fome Zero (Zero Hunger), started to transform this history. In the 1990s, the state conducted a reform that positioned Ceará as a "case study" of successful water modernization in the international arena. Shortly after the democratization of Brazil, and with the support of the World Bank, that reform overhauled the state's legal architecture and its water administration system. The state increased its capacity to store and move water with new reservoirs and established programs to generate more scientific information about the relation between climate, water, and agricultural production. The reforms also introduced a participatory reservoir management program, which, paired with a payment system for bulk water, emphasized water's commodified social life. Water users, those with demonstrable financial and legal claims over water, entered into a system of rights and payments that entitled them to participate in making coordinated decisions about water allocation. Managed as a collective resource, while mediated by its commodification, water led a liberal political life (Ballestero 2004, 2006).[2]

Given that Brazil was one of the few Latin American countries in the United Nations system that at the time did not promote the recognition of the human right to water, those reforms, especially the charge system, were described by water professionals in Ceará as an ethical act. This idea was not totally unfounded if one considers the historical appropriation of water by large landowners and agribusinesses, who privately benefited from state investments in water infrastructure. The idea of charging those landowners and businesses for water resonated with many as a distributive justice measure.[3] Commodification was a political tactic aimed at raising financial resources to make water universally accessible, as its recognition as a human right requires. In this way, commodification was entangled with its humanitarian distribution.

As I mentioned before, Ceará's 1990s water reforms were broad. They did more than reorganize public agencies and ministries. The reforms reimagined the economic potential of the state's semiarid environment by promoting more "efficient" water use, such as irrigation for fruit production for export to Europe and the United States. This led to the creation of new irrigation districts at the expense of smaller-scale rice farmers who had been historically supported by the state. Water managers also began incorporating meteorological data and climatic predictions in their decision making. Politicians proclaimed these measures were an economic rebirth for Ceará, and technocrats focused on the details of how the new water rights system, participatory decision-making structures, and charging mechanism would work. The charging model they designed was difficult to implement, and to date only a fraction of water users actually pay the charges (de Oliveira 2008). The water rights and participatory management systems, however, were relatively successful, even if they were unable to completely turn around the legacy of inequity and water scarcity that shaped people's everyday lives. Ceará became a poster child for cutting-edge water management.

At the beginning of the twenty-first century, proponents of the Water Pact were convinced that, despite the 1990s reforms, clientelism and the legacy of coronelismo still impeded the securing of water access for all. Pact organizers believed a new approach was needed. They contended that moving away from predefined engineering or legal solutions and tapping into people's moral sensibilities would create a new reservoir of energy and resources that promised to yield better results. They wanted to change Ceará's future history by drawing on people's sense of everyday ethical action. To do so, they would not make the mistake of beginning with water infrastructure itself—determining ways to move and allocate water—nor would they turn to the law and its universal obligations. Instead, they would focus on people's inherent capacity to care, making every member of society recognize their moral obligation to *cuidar da água* (care for water) according to their own capacities and in harmony with their everyday contextual demands. This turn to morality is not unique to the pact. At the end of the twentieth century, morality became central in neoliberal forms of governance, where responsibility for social care was relocated from public institutions into the intimacy of citizens' everyday lives (Muehlebach 2012). Relying upon individual philanthropy, nongovernmental organizations, religious assistance in emergency situations, and old and tried gendered divisions of labor, the state stopped claiming as its responsibility the provision of welfare, health, and care for children and the elderly. The pact is clearly inspired by this shift in the allocation of responsibility over social and common concerns, and yet something else was also going on through it.

UNDOING A HISTORICAL MISTAKE

As I looked out the window of the car taking us to the WP meeting in the Carirí region of Ceará, a continuous layer of dry shrubs, no more than three meters high, surrounded us on both sides. The vegetation was dry, thorny, and striking for someone like me who is used to the lush greenness of Costa Rica's tropical rain forest. I was seeing more shades of brown and gray than I had ever seen before. As the sun rose, the moving images were breathtaking. The *caatinga*—the biome found in this part of the world—has an unusual capacity to scratch your senses. During the dry season its sharp angles and denuded sticks leave no doubt about what life with limited water looks like. I was in the car with Zé María, a Water Company driver, and

Flavio, who was sitting in the passenger's seat. I noticed he studied my facial expressions through the side mirror, as if amused by my fascination. He then made a shocking statement: "What you see out the window is a historical mistake [*um erro histórico*]. People should not live here."

I was taken aback by Flavio's words. I had known him for a while. He was the director of the local watershed management office and a veteran of water institutions, having once served as the statewide director of the water management company. Flavio has big ideas, many of which do not resonate with his colleagues. Despite this, he is well respected. Community organizations and state officials give him space to experiment and implement new projects, even if they sometimes disagree with his views. While not using the language of historical mistakes that Flavio uses, most people in Ceará share the sentiment. People have a sense that, for whatever reason, they live in a difficult environment, an area that constantly tests the ability of life to reproduce itself. This sense is a source of both pride and pain.

As I looked out the window I could see what Flavio meant: extremely poor communities punctuating the flow of the caatinga; families without the possibility of accessing any water for their subsistence agriculture and barely covering their physiological needs. While I could see what his comment implied, characterizing as a mistake the wealth of human and nonhuman relations that fill days and nights of the sertão was dangerous. Too many colonial and imperial echoes emanated from the statement. But Flavio was not naïve in any way, nor was he proposing to erase the multiple forms of life unfolding in the caatinga. He was clear that this "historical mistake" was not the work of "nature," but a human error that needed to be righted. He also was clear that it was the state's responsibility to do so.

The idea that people living in the sertão is a historical mistake is relatively widespread inside and outside of Ceará. The problem of how to bring water to the *população difusa* (diffuse population) has persisted for centuries. The term diffuse population refers to small villages dispersed throughout a large swath of the state's territory (146,348.3 km^2) and disconnected from water infrastructure. When state officials contemplate the goal of providing access to water for all in Ceará, they are thinking about diffuse populations. The difficulties diffuse populations face in making a living have resulted in cyclical waves of migration from the hinterlands to urban centers in the state and to the industrial, financial, and political centers of the country in the south and southeastern parts of Brazil. In those regions, stereotypes paint *nordestinos* (people from the northeast, including Ceará) as poor and uneducated, trapped between the horrors of *seca* (drought) and their own backwardness.[4] During the 2018 presidential election, the extreme right wing, evangelical candidate Jair Bolsonaro, and his followers, used a stereotypical view of nordestinos as a lynchpin of his extreme and neo-fascist propaganda. While Bolsonaro was ultimately elected, he did not win any of the 184 municipalities in Ceará. Moreover, the northeastern region as a whole voted for the leftist Partido dos Trabalhadores.

Made vulnerable to the politics of water scarcity by decades of land concentration among political elites and now multinational corporations, residents of Ceará have seen a parade of ideas march through their neighborhoods and villages. Ranging from cloud-seeding technologies to underground household water tanks, all of these ideas have been announced as definitive solutions to people's exclusion from basic water infrastructure. None of them have successfully led to the transformation that politicians and state officials promised. But despite their recurrent failures to secure universal access to water for all of its citizens, and regardless of its reputation as a "backward" state in other realms, Ceará is also deemed an example of good water governance by specialized international water institutions (do Amaral Filho 2003; Simpson 2003; Tendler 1997). This recognition came after the 1990s reforms.

And yet, at the beginning of the twenty-first century, a vast portion of the population in Ceará continued to buy water in bulk from "horse-drawn carts, motorized tanks, or from people who walk around the streets with large cans of water" (Caprara et al. 2009: 128). In the rural areas, vulnerability to cyclical droughts continued to make households dependent on emergency measures. Notwithstanding the inauguration in 2002 of the Açude Castanhão (Castanhão reservoir), the largest multiple-use reservoir in Latin America, water scarcity continued to lurk in people's memories and shape their everyday life. In 2013, for instance, the water level of the Açude Castanhão was so low that the ruins of Jaguaribara, the town flooded to create the reservoir, emerged as evidence of the horrors and failures of large-scale infrastructural solutions to water problems. By November 2017, after six years of extreme drought, the Açude Castanhão was down to 3 percent of its storage capacity, reaching its *volume morto* (dead volume)—a water level so low that the reservoir was declared no longer functional.

Aware that the results of the previous reforms were mixed at best, representatives in Ceará's legislature turned their eyes again to water. The state had not only been unable to resolve its dramatic water access gaps completely, but was also losing its national and international edge as an innovator on water issues. After much discussion about the practicality of passing new water laws or reforming institutions, the representatives amplified their discussions to query what kind of political action was needed. They agreed that the difficulties in the water sector were symptoms of something else: the legislature had lost touch with society, its communication channels were broken, and it needed a more dynamic way to understand the needs and difficulties people experienced.

As a result, in 2007 the legislature officially created the Council for Advanced Studies and Strategic Issues. The body was tasked with reducing the "distance that separates citizens from their congressional representatives," identifying strategic statewide challenges, and producing recommendations on how to confront those challenges. To achieve the social improvements they hoped for, legislators had to stop acting as if liberal political representation was a natural vector of the social. The idea that some people can represent the needs and interests of others was no longer taken for granted; put differently, people were less tolerant of the idea that political representation was a transparent act. Congressional representatives sought a new way to connect with society.

Not surprisingly, when the council considered possible foci of action, water quickly emerged as a top concern. The council's personnel settled on the idea of producing a statewide Water Pact, something that had never been attempted before in Ceará or Brazil and, as the director of the council told me, something that was going to become a hallmark of collective policymaking for the twenty-first century. Against this backdrop, the council worked for approximately four years (2007–10) to construct the Water Pact as a form of political mobilization that could right the historical wrong Flavio had identified as we looked out the car windows.

Although formally presented as an effort led by the legislature, the pact was really the responsibility of the council and its personnel. Ernesto, a seasoned public figure in Ceará and Brazil more broadly, was appointed executive director of the council.[5] Years before, Ernesto had famously resigned his position as state secretary when he was pressured to act unethically in one of the water programs he oversaw. Having served in multiple positions, he had the necessary political experience and ethical standing. He selected a group of consultants on the basis of their commitment to water-related technical activism to help him design and implement the pact. The team came from academia, NGOs, and public institutions, and combined diverse areas of knowledge and experience: engineering, hydrology, sociology, and geography. They shared a common vision. The pact had to be different from previous "technical fixes," where authorities reserved ultimate planning and decision-making powers for themselves. These technical plans had become elusive promises, characterized by "failures, mismatches, discrepancies and gaps" (Abram and Weszkalnys 2013: 22). While the consulting team held sophisticated views on water issues, they wanted to privilege people's personal context and avoid fully imprinting their ideas on the pact's content. In Ernesto's view, the methodology of the pact could be predesigned, but its content could not; it had to be locally determined.

Considering this commitment to people's context, Ernesto and his consultants designed a method that would produce not one, but a series of pacts—one for each municipality, alliance of municipalities, watershed, and ultimately the state as a whole. But these scales, while nested, were not hierarchically ordered. The commitments made at the state level, for example, did not overrule those made at the municipal level. The strategy was to affirm the independent coexistence of all the pacts, to make promises proliferate, rather than centralizing relations of obligation and debt on mayors, the heads of the water agencies, or the state governor. All promises made through the pact would retain their forceful existence, affirming the particular contexts from which they emerged.

The method of promise proliferation the consultants settled upon was the public meeting. These meetings were held in school classrooms, public auditoriums, and municipal institutions with spaces large enough to host, sometimes, more than one hundred people. In order to attract attendance, the consulting group sent invitations signed by the president of the legislature, and in some cases, by the governor of Ceará. They also spent considerable time on the telephone, motivating key actors in a variety of public institutions (municipal, state, and federal) so that they, in turn, would encourage people in their offices and communities to attend the events. A day or two before each event took place, teams of consultants mobilized in the locations where the meetings were held. During this time, they paid visits to the local offices of institutions and organizations from which they had not been able to secure participation. In the visits I was part of, we talked to Catholic Church representatives, technical institute educators, rural agricultural extension officers, and regional health supervisors. Some days, the team conducted up to eight such meetings, using each visit as a way to secure personal support through the obligations that embodied interactions create.

The day of the pact-making event, local institutions provided transportation to bring participants to the municipal centers. Alternatively, people activated their own networks and organized collective transportation in cars, motorcycles, and minibuses. At the event, time was used according to a set of steps the consulting team had disseminated in an electronic document that also outlined the expected results. The pact consisted of the promises people made, their documentation, transcription, and final aggregation into at least one pact (municipal, watershed, or state level).

BUILDING MULTIPLICITY

Along with his consulting team, Ernesto also hired a permanent staff consisting of a secretary (his own lifelong personal assistant), an information technology specialist, and a journalist in charge of public relations and communications. To launch the pact, he raised federal and state funds to hire a group of twelve consultants—although the number would increase and decrease throughout the project. This group gave shape to the pact on the basis of Ernesto's own personal interest in related international projects such as the Zaragoza Water Charter, the European Water Directive, and the work of the Foundation for a New Water Culture.

Rebecca was one of the consultants Ernesto hired, probably the most experienced in the group. Trained as a sociologist, Rebecca has been involved in water and social justice projects for all of her career. She was a central participant in Ceará's 1990s reforms and had previously worked for Ernesto on land reform projects he launched while serving as agriculture secretary years earlier. A renowned professional, Rebecca is hired across Brazil for her expertise in participatory policy processes. She has a strong presence and soft manners when not upset. Known for being straightforward with her interlocutors, regardless of their political standing, she was key in the design of the pact. When we met in her apartment, she welcomed me with the obligatory *cafezinho* (coffee) people offer you when visiting a household or office.

That first conversation took the whole afternoon. It was clear that for Rebecca the pact was the most interesting political experiment at the time. She was energized by it, largely because she felt that things could really turn out differently this time. Instead of reproducing trite policy models, they were entering uncharted territory by building a pact and not a more familiar law, policy, or plan. She explained this political, technical, and affective juncture by saying, "[In the 1990s] we did it all, we applied the *pacotinho* [the little package] of policies just as the international establishment recommended. Our state is a textbook example, and what happened? Fifteen years later we have more *carros pipa* [water trucks] than before. So we needed to do something different. And this is where the pact emerged. What is new these days in Ceará is the Water Pact."

To a large degree, this disillusionment with specific prescriptions to solve water problems was responsible for the pact organizers' devotion to doing something different, to taking a new approach. When I asked Rebecca what exactly the pact was, she answered by noting the difficulty of articulating a definition: "The pact is not a thing or a government plan, but a framework that will survive governmental changes and effect more perennial commitments. It is a way of doing things. It will go beyond political preferences to determine *caminhos* [paths] to be followed regardless of who comes to power."

I was surprised by her answer. Ernesto often mentioned that the pact would produce something like a strategic plan, and yet one of the central figures in the process, Rebecca, refused to bind the pact. She did not want to turn it into a familiar form, although her invocation of a path was telling. In a state where lasting commitments are embodied in concrete infrastructures such as canals, reservoirs, and irrigation districts, her projection of the pact into the future necessitated infrastructural analogies. But contrary to infrastructural projects, where people converge on a reservoir or canal as a possible solution to a problem, even if that structure means different things and is used in various ways by different groups, the pact was more ambiguous. Rebecca's idea of a path was not a singular proposition. Rather than emphasizing how a path delineates a route, her analogy emphasized something else.

When following a path, Rebecca explained, if you raise your eyes from the ground, you see all sorts of interesting, diverse, and surprising things. It was the capacity of a path to reveal multiple things that Rebecca aligned with the pact. She noted how the lack of a shared definition of the pact was irrelevant. For her, it was not important to ask what the pact was; what mattered was figuring out how to mobilize people's capacity to think from their particularities rather than focus on the state and the funds it could transfer. If the pact was going to leave any lasting mark, it would not be because it proposed a singular solution but because it kept alive the multiplicity that the method of its construction was designed to yield (on method, see Miyazaki 2004).

The pact's emphasis on how, instead of what, marked another important difference with previous efforts to deal with the material politics of water. Rather than focusing on finding a shared definitional space, as when people attempt to settle on a "solution" to a particular water problem, the pact team worked to reorient people's everyday activities, whatever they were. That is, rather than designing exceptional interventions centralized in government agencies, the pact would produce a quantity of responses that could not be controlled by the "usual suspects": politicians and economic elites.

This is also the reason why the pact team did not want to begin by defining what needed to be done, an exercise that was often co-opted by technical vocabularies and political cliques. They hoped to show participants how to arrive at their own ideas of what needed to be done in their localities. By redirecting their focus from *what* collective actions were appropriate to take, and instead focusing on *how* to invite people to act (Mokyr 2001; Ryle 1945), the pact team asked bureaucrats and the public to engage the slippery everydayness of their relationships with water.

The pact organizers believed that this methodological choice would prevent the pact from becoming a list of technical solutions, accumulated as if they were possessions of a few social actors with the requisite amount of capital and technical knowledge to implement them. They wanted the pact to be about different flows of action; dynamic arrangements of bodies, affects, and materials; minor rearrangements that when aggregated could yield major transformations. Regardless of your position, you could be part of the pact. Thus, the promises people made took a seemingly unremarkable form: ordinary, everyday work commitments. Here, the intimacy of care was the intimacy of one's daily obligations. Take the following examples of promises made in a municipal pact:

- I can make sure that we include environmental education programs in our municipality.

- I will fight to get the resources to expand the *adutora* [water main] the municipality has been planning to build for three years.
- I will lobby my fellow health workers at the regional health directorate to start talking about water conservation with patients.
- I will hurry up the training program we have on the books to share information about more efficient irrigation technologies with farmers.

As we can see, the promises in the pact have distinct material implications, involve different people, and activate different social relations. Their effects could only be recognized at that level of specificity. In this sense, the pact sought unruly multiplicity, myriad promises impossible to measure, narrate, verify, or uphold according to a single standard. Not many promises of one kind, but many kinds of promises.

This complex structure explains the long hours the pact team had to spend determining the methodology they would follow. Every Monday afternoon throughout the approximately four years that the pact was being built, the team met for three to four hours to discuss their methods. In a meeting room located in the premises of Universidade do Parlamento, the Parliamentary University (a training space where congressional representatives and their staff came to increase their knowledge about all sorts of topics, from management to legal theory), the consultants reviewed their recent activities, adapted their methods to new knowledge they acquired, and planned future activities. Those meetings were attended by the consulting team and ranged from five to fifteen participants, depending on the number of consultants active at any particular point. Considering how new the pact was, and given the lack of any precedents to emulate, that meeting room witnessed hours and hours of reflexive discussions.

A recurring topic at the Monday meetings was unruly multiplicity and the extent to which it was a political innovation. Multiplicity and contradiction were not new for people in the room, or for Cearenses for that matter. If anything, the history of water politics the pact team wanted to break with had had too much contradiction and unruliness. Whether in the form of a powerful rumor, a suspicion of a back-room deal, evidence of "corruption," lack of discipline among public officials, or citizen disregard of legal prescriptions, multiplicity and unruliness were everywhere in Ceará. As had happened many times, a meeting to decide on the amount of water to be discharged from a community reservoir could very well become all about the accuracy of a measurement instrument (Ballestero 2012). But in those instances, unruly multiplicity was a problem, an obstacle to a predetermined path intended to take technocrats from point A to point B without distraction.

In the pact, however, unruly multiplicity had a radically different value. If in the past state officials had dealt with unruly multiplicity as an unintended consequence, this time it would be front and center—an intentionally embraced condition of social and material life. To generate the right type of unruliness within the constraints of a technocratic effort, the pact team established a set of principles to provide enough structure so that multiplicity could be generated without unleashing absolute chaos. The first principle stated that the pact could not *atropelar o sistema* (run over the system), referring to the existing water management system put in place during the 1990s. Their aim of breaking with history was shaped by what already existed, even if it was supposed to transform it. The pact team did not presume the world was an empty slate waiting to be marked by their latest ideas. Any changes the pact introduced would be shaped by the legal and administrative institutions already in place. The system had to be changed, but also kept as it was.

The second principle stated that the pact would strive to generate *compromissos reais* (real commitments), inescapable obligations. Previous policies, they thought, had a singular focus on law that did not speak to the "reality" of people's everyday and intimate experiences. The pact would tap into that reality by engaging everyday life, a space beyond the reach of formal legal obligations. And finally, the third principle reminded them that if those compromissos were to resist the turbulence of electoral cycles, they would have to tap into people's sense of moral responsibility; the pact had to make people care (*cuidar*) for water in whatever way they could and according to their context, resources, and jobs. This kind of professional and public intimacy did not align with the domestic, the bodily, or the emotional. It was openly policy-related and technocratically shaped.

Following these principles, Rebecca explained to me, would yield a quantity of social energy that had no precedent. It would create an arrangement with enough temporal endurance (beyond electoral cycles) and moral texture (by creating inescapable personal obligations) to break with history (without ignoring its hold on the present) and come close to the dream of fulfilling people's fundamental right to water in an ethical way. The pact was not a modernist proposition that assumed anything was possible. Breaking with history implied attending to its stickiness.

FROM MONUMENTS OF CONCRETE TO BUREAUCRATIC CARE

The responses of public institutions to water problems in Ceará have traditionally followed what a water agency official referred to as *a política do concreto* (the politics of concrete), a concept that signals an almost automatic turn to physical infrastructure to deal with water scarcity issues.[6] Although explained by some technical personnel as a transparent technoscientific fact, water scarcity has been historically understood by everyday citizens in Ceará as a problem caused not only by environmental and climatological conditions, but more importantly by the political trade of waterworks between coroneis, landowners, and regional authorities (Ballestero 2006).

People in Ceará already have a deep awareness of the liveliness of the material world, including water infrastructure, and its mutual constitution with sociopolitical processes. They easily recognize water infrastructure projects as political mechanisms that create subject positions, relations of political debt, and access to or exclusion from state benefits. In Ceará, water has never been a matter of fact and instead has always been a matter of concern whose political and social entanglements could never be reduced to the technical, despite the efforts of some actors to do so (cf. Latour 2005). Aware of this, the WP team's intent was not to depoliticize or depersonalize water through technical knowledge. Quite the contrary, they were looking for a path through which water could be turned into an inescapable and intimate political responsibility. They wanted to go from the strategic manipulation of reservoirs and canals to an indisputable collective responsibility distributed among all public actors. Ultimately, the pact was intended to revalue quotidian, nonmonumental care as a more durable form of concern in water politics.

Doubts about how to promote care through bureaucratic design emerged many times during the team's Monday planning meetings. One afternoon the group was discussing what guidelines they would provide to the organizers of municipal pact meetings. They were discussing the type of questions they should send to guarantee that a "different" outcome would come out of a municipal pact:

ERNESTO: I don't want us to pose questions that are closed and rigid. If we ask those questions, we end up with homogeneous answers from municipalities that are radically different. So we must think carefully what are we going to ask from them. We need to make them think close to home.

LUIS: We have to convey that the challenges we are working on are not problems. We have to frame them as positive statements with which the municipalities can identify. The way they are now is too abstract. If you ask about the water management system, they will respond with requests to Fortaleza for infrastructure, and we have to avoid that.

ERNESTO: Our ultimate goal is for the *município* to see itself in the document, but also to take a position. To commit to doing something, right there, without asking somebody else to do it.

LUIS: The questions have to detonate discussion and commitments.

REBECCA: We need something that takes the discussion away from infrastructure. Something that encourages them to see their obligation to act today instead of asking someone else to build a new structure.

FERNANDO: But I really don't think any commitments are going to come out of the municipal meetings. The real commitments will be the result of the regional meetings.

ERNESTO: We don't have to be scared. This is not going to be one more technocratic document. This will be a document with a future-oriented perspective. We are not going to repeat the model of asking the state for everything. That is not possible any longer. It is stated in the Zaragoza Charter, the moral responsibility for water is shared.

Besides showing the ambivalences among team members, this exchange highlights two things. First, it shows how the pact is inscribed in broader, global shifts in people's understanding of the role of the state that we might gloss as neoliberal. The state could no longer respond to all the demands people placed on it—as if it ever did. Second, the conversation also shows the intricate correspondences between care and context. Rather than moving to larger administrative jurisdictions, the team understood care as "doing something close to home." Ernesto's opposition to homogenization was an attempt to foster an appreciation of the multiplicity of everyday labor that goes into life and politics. This revalorizing the specificity of one's context, rather than its homogenization through legal means, resonates in an uncanny way with feminist concerns over care as an "ethico-political commitment[s] to neglected things, and the affective remaking of daily life" (de la Bellacasa 2011: 101). Another of the consultants articulated this relation between care and context in even more blunt terms when he said, "For the pact to really take place, its construction needs to be collective. The person participating needs to see some of her own desires in the pact, her expectations, her immediate surroundings, her ideals, right? In order for people to be willing to fulfill their commitments, I believe a fundamental aspect is that you can see yourself in that agreement. See your willingness to work, your expectations, your context, those kinds of things. That is the only way people will care today to change the future, even considering the difficulties they face in their everyday actions."

This caring is, of course, inscribed in hierarchies of power, bureaucratic legacies, and histories of capital accumulation. After all, Ernesto's and the team's convictions about the need to involve society were a variation of ideologies of shared public–private responsibility that, paradoxically, they also criticize. So this form of care is not a romantic endeavor devoid of interests and betrayals. Yet the organizers still imagine the care for water that the WP encourages as an effort to think how things could be different in a future history that is engendered through the specificities of people's here-and-now. Repersonalizing water by valuing the immediacy of everyday action, as opposed to centralized grand political programs, was the change the organizers sought to make public officials embrace their moral responsibility to care and commit to act.

EXCESSIVE RELATIONALITY

The pact's commitment to promote care for water was grounded in a robust network of water institutions that include the water resources secretariat, the water management company (WMC), the water service provision company, a water emergency management system that is activated during times of extreme drought, a system of community aqueducts, a network of municipal water management offices, and myriad water-related programs in the education, agriculture, and health secretariats. These institutions think with and through water as an abstract substance as much as they treat, move, and distribute the liquid. They deal in formalities and abstractions as much as in substance and matter.

Amid that dense institutional landscape, the WMC stands out as one of the most technologically advanced and innovative agencies. The WMC was created in the 1990s as a private company owned by the state, by its workers, and by a small group of private shareholders. Despite its private nature, the WMC has a peculiar identity that comes out in everyday conversation. Almost every person I met at the agency emphasized their responsibilities as public servants, even though technically the agency is formally private. In internal and public meetings, WMC employees refer to their ethical responsibilities as part of the state; they repeatedly mention things like their obligation to be responsive to public demands and to lead when society lacks knowledge about water issues.

When the pact began to take form in 2007, top officials at the WMC committed to support it. The directors of the WMC see any water-related initiative as an opportunity to spread knowledge and collect information about water management. Moreover, they believe that the WMC has to be involved in any water initiative launched in the state. As the agency in charge of bulk water management—it is responsible for reservoirs and for bringing water toward its users—WMC employees see themselves as the organization closest to the origin, the organization responsible for furnishing water for society. The WMC president and head of planning often attended the Monday pact meetings in Fortaleza. The two men are well-known figures in Ceará's water politics. One became federal secretary of water resources under President Lula da Silva and the other replaced him as president of WMC. At the time of the pact, their attendance was a symbol of WMC's support.

Despite that symbolic support, some técnicos at WMC were more cautious. José and Pedro are two of those técnicos. Since they started working in WMC, both have steadily moved up in the company's hierarchy. They are well-known for being critical thinkers and good managers, the kind of people who know how to make things happen and express their critical perspectives directly. They are also known as strict leftists, often expressing their party allegiances—they were both members of the Partido dos Trabalhadores, the party of the then president Lula da Silva—and openly criticized the heavy-handed intervention of multilateral financial institutions in domestic matters. Early on in the creation of the pact, they expressed their doubts about the whole process.

One weekend, I caught a ride with José and Pedro back to Fortaleza after a pact meeting in the Jaguaribe watershed. I had already heard about their opinions secondhand, so I directly broached the topic once we were on the BR-116 highway. Our conversation was lively. José and Pedro were not very excited about the pact. They feared its outcomes could weaken the water management system they had spent so much time, more than a decade, putting in place. But they were not defending the indefensible. They knew that WMC's management system needed to be improved. From their point of view, one of the main challenges they faced was making sure that "society" understood the role that WMC played. The WMC did not occupy as prominent a place in people's imaginary of the public life of water. As it turns out, everyday citizens were much more aware of what municipalities, health and education secretariats, and even the military did during times of water emergency. The pact, José and Pedro believed, would reinforce the prominence of other institutions over the work that WMC was doing.

Beyond tactical disputes over political and institutional prestige, José and Pedro had another reason for not believing in the pact. "Society did not ask for it," Pedro told me. Nobody, except for a small group of politicians and technocrats, had asked for anything like the pact. Pedro was drawing my attention to the fact that the pact was not the result of any process through which people could communicate their collective need. I was puzzled by Pedro's comment. As far as I knew, neither the 1990s reform that gave birth to the WMC nor any of the smaller projects that followed it had resulted from grassroots, or as they said "societal," demands.

I later realized that the problem Pedro was diagnosing was less about society and more about the fact that the impetus for the pact originated in a single location, from a very small group of people. The idea and its implementation originally came from Ernesto and his consultants. This origin brought the initiative dangerously close to the legacy of coronelismo and to personalized water relations that WMC had worked so hard to interrupt. From Pedro's point of view, rather than a break, the pact was a continuation of the history of personalization and commodification of water through relations of political debt. The pact was too personal. Pedro was right about this, but perhaps contrary to what he assumed, that was not an oversight. Ernesto and his team wanted to intensify and take advantage of this personalization. They wanted to work with it rather than against it. The pact was not going to attempt to undo the personalization and intimacy of water relations in Ceará technocratically, the way the WMC wanted to. The pact was going to mobilize the potential of the moral power of people's relations of debt and obligation (see also Ansell 2014); it was going to expect more from personalized links. It was not going to try to extinguish them.

Yet for José this personalization represented a weakness, a likely pitfall. He explained, "if you need the governor to call secretariats so that people attend the pact meetings, that means the process is not mature. There is no felt need. The mobilization is not really working." The phone call José used as an example stood for the intricate and steady labor that Ernesto, the diputados (congressional representatives), mayors, and consultants did to motivate people to participate in the pact meetings. The pact team invested a lot of time monitoring the level of excitement about the pact. Before and after each regional meeting, for example, members of the consulting team called all mayors to secure their recommitment to the process. Project assistants or secretaries organized the calls, and whenever possible Ernesto spoke personally with them. Mobilizing Ernesto's charisma as a prominent political figure in Ceará would prevent the excitement for the pact from cooling down. If conducted successfully, the calls culminated with a promise, a commitment to be part of the pact and to enroll as many people as possible by making their own calls and paying their own visits. These promises were forms of reciprocity for previously established obligations or newly created ones in anticipation of future events. Society was ultimately the result of the density of all those relations of political debt among public servants.

For Pedro and José, the problem was that by relying on these intimate relations the pact asserted a form of "excessive relationality" that, in their view, Ceará really needed to move away from. The legacy of this excessive relationality could be seen all around the state in the personal debts that ruled public life. That relationality was responsible for entwining governmental interventions, public drought relief resources, and public investments with personal debts among politicians, landowners, and residents. That relationality is also well understood by people in Ceará. Scholars describe it as relations of patronage and exchange of political regard. Pedro, José, the members of the pact's consulting team, and Ceará's residents more broadly refer to these networks of camaraderie and resource redistribution as the drought industry (*indústria da seca*), a system that reaps benefits from drought relief efforts via clientelism. In this system, water has been commodified as the currency on which influence can be exchanged. For Pedro and José, the pact—with its dependence on personal phone calls, office visits, and handshakes—was too close to that history.

Based on their reading of Ceará's history, José and Pedro predicted the pact would not be successful. Turning water into a human right by undoing its commodified role in the drought industry was a complicated task that the pact most likely would not achieve. And yet, like many, despite their forecast, Pedro and José had an institutional commitment they were going to fulfill. WMC had pledged to be a central player in the pact, so they went on to help organize the promise-making rituals. But they did not see a very compelling future for the pact. To an extent, they were right in their predictions. Once the pact concluded, its effects were uncertain, fuzzy, almost as if the pact itself had not achieved a full existence.

DIAGNOSIS: MAKE ROOM FOR CONTRADICTION

In late 2008, I attended a ceremony at Ceará's Assembléia Legislativa (Legislative Assembly). I arrived at the building after walking past a couple of carts renting "appropriate attire"—ties, jackets, dresses—for people who wanted to visit their representatives and make the right impression. The Assembléia building has an architectural style reminiscent of a 1980s science fiction film: it looks like a flying saucer sitting on a green field, separated by a fence from the buzzing streets that surround it. I entered the building through a side door that funneled me and all the other visitors toward another door that opened onto the floor where congressional representatives hold their debates.

On that day, Ernesto was scheduled to present to the state's water expert community a diagnosis of water issues in Ceará. The state governor, Catholic Church authorities, water and agriculture secretaries, and a host of state and federal political figures filled the main table. The speakers were flanked by two flags, to the left Brazil's and to the right the state of Ceará's. Between them a large cross hung from the wall overseeing the discussions regularly held by members of the Assembléia. Ernesto sat at the center of the long table, right under the cross. Once his turn came to address the audience, he began by explaining the steps the pact team had taken thus far. He announced that the first phase of the pact had concluded with the publication of the document he was there to introduce. Once he began describing the document, he spoke of the report as a technical document unlike any other in Ceará's history. He told his audience, "ninety-seven institutions and more than five hundred hands, from two hundred and fifty six técnicos [technical personnel]" had written it. Raising the publication into the air and using his arm as a pivot to make sure everybody could see its green cover, he proudly described the report as a "document full of contradictions" designed "not merely to produce a picture but to interpret that picture."

At that point, I saw more clearly how ideas of unruly multiplicity seeped through the pact. This was more than mere tolerance of the well-known fact that people hold different ideas that can be in contradiction with one another. Ernesto believed that contradiction was good, something to be promoted, an index of the pact's future success; if the pact housed deep contradictions, then they had been effective in capturing the diverging conditions under which people live their lives and deal with water. After approximately ten minutes of speaking, Ernesto changed his voice, signaling that he was coming to an end. He slowed down and said, "Just like Paulo Freire asks us to do, this document is not a *mirar* [look], but an *ad-mirar*, which is to look inside with a critical perspective."

For the critical theorist and education scholar Paulo Freire, ad-mirar is the act of creating knowledge by introducing distance between that which is to be known and our already existing ideas about it (Escobar 1972: 24). Freire viewed knowledge as a resource for political transformation; for him ad-mirar is an act that requires questioning the epistemic and ideological obstacles that preclude people from critically perceiving the world around them and modifying it. The invocation of Freire's ideas, an important figure for leftist and popular education movements in Brazil and Latin America more generally, gave Ernesto's speech a poetic and idealistic tone.

After quoting Freire, Ernesto paused briefly to gauge the effect his invocation had had on the audience. He then thanked the audience for their attention, opening the door for a long round of applause. As he stepped down from the elevated stage where he had delivered his speech, he was immediately surrounded by técnicos, NGO leaders, mid-level managers, and federal authorities who had traveled to Fortaleza for the event. They shook his hand, took pictures with him, found ways to congratulate him on the document, and reminded him of the promises he had made in the past or support he had offered for particular projects.

Ernesto's invitation in that day's speech to engage in critical analysis and to let contradiction proliferate was prophetic. Or, seen from a different perspective, it was a display of how attuned he was to the asymmetries behind Ceará's water history. Contradiction was not only an epistemic and political aim of the pact; it was palpable in the personal encounters that would produce the ad-mirar Ernesto sought to foster. Contradictory life experiences, skills, aspirations, and capacities structured the encounters that were at the basis of modifying the world of water politics with its clientelism, economic exploitations, and cyclical institutional reinventions.

It was clear that Ernesto's attention to contradictions was not an abstraction. Experience had shown him that any attempted solution to the water crisis, even if adopted by consensus, inevitably generated conflicting effects. For example, the construction of Ceará's largest reservoir, the Açude Castanhão, turned surrounding communities into illegal water users as they suddenly were required to secure a permit to use water for their farms. Another example was the bulk charge system that, despite being intended as a distributive justice measure, ignited a fierce debate about the justice of a waiver the law established for subsistence farmers consuming small quantities of water. As with any program or policy, with these interventions some people inevitably benefited and some people were left worse off. This history made Ernesto unafraid of contradiction. On the contrary, he knew that the only thing left to do was to let contradictions proliferate without trying to fully align people's interests, social relations, and conflicts.

THE HANDSHAKE THAT NEVER WAS

With the publication of the diagnostic document, the WP moved into a new phase that consisted of public meetings where the promises, the substance of the pact, would be made. These meetings fostered innumerable encounters among state, municipal, Church, and community organization representatives. Some of them were dramatic, others more subdued. But in all of the meetings, to one degree or another, differences between the participants shaped the interactions and the outcomes.

A couple months after Ernesto's presentation, I was at a municipal WP meeting in Limoeiro do Norte, a town of about fifty-five thousand inhabitants that had become a "bedroom community" for workers of the agribusinesses who came to the region in the early 2000s to take advantage of tax breaks and a new irrigated perimeter built by the state. When the meeting turned to a collective discussion, the consultant responsible for facilitating the event found himself in an argument with Antonio, a municipal worker about forty years old, who was known for his union activism and for his ability to turn public events into performances of his political influence.

While sharing the stage, Antonio challenged the consultant, noting that water scarcity issues were as old as the town itself. The consultant was a French agricultural engineer who had recently received his PhD in a joint program between a French and a Brazilian university. He came to know Ceará very well through his dissertation research, which focused on the potential of shallow wells as water sources for rural populations. The consultant had been involved with the pact from the beginning, having gained Ernesto's trust early on. During the Monday planning meetings their opinions often aligned, constituting a powerful rhetorical front that other consultants seldom challenged.

A few minutes into his onstage exchange with Antonio, the consultant was explaining the difference between a pact, as a political project, and a conventional law, like the ones the state had produced in the 1990s. The large auditorium was only half full, but the audience was engaged. In the middle of his speech on commitments, promises, and responsibilities, the consultant extended his hand into the air, offering it to Antonio, inviting him to perform the act of handshaking that seals a pact as a promise to one another. With his arm extended and his hand waiting to feel the touch of Antonio's, while still facing the audience, the consultant continued speaking on the importance of keeping one's word, and he announced to the audience how the public nature of this pact, and the honor one puts into a promise, would discourage pact makers from breaking their pledges. After what felt like an interminable wait, hand extended into the air, waiting for his interlocutor to make contact, the consultant finally turned his head searching for Antonio only to discover that he had taken a step away from him, increasing the distance that separated them. Antonio had also interlaced his hands behind his back and was offering nothing but a smile. He knew his refusal was a breach, but he did not soothe the tension generated by his interruption. After some nervous laughter from the audience, the consultant also smiled and brought his hand back to safety, close to his own body, and continued with his presentation. Antonio returned to his seat and the meeting continued according to schedule: group work, promise making, and public presentation of the results to the rest of the participants.

This handshake refusal indexes the intimate and conflict-ridden encounters through which contradiction and promise-making became "new" ways of dealing with water problems.[7] But it would be a mistake to interpret the unconsummated handshake as a symptom of the failure of the pact. It is true that the municipal worker's refusal to shake the hand of the consultant was a dramatic act, a powerful reminder of the impact small breaches can have in everyday relations between social actors with radically different histories. And yet, in the days that followed the meeting, the municipal worker became an active participant of the pact. He helped organize local meetings, rallied for participation among local critics, lobbied other politicians to get involved, and demanded from the organizers more attention to ongoing local political events. Antonio became an enthusiastic promoter of the pact while also reminding people about his deliberate refusal, enacting his reservations about being fully incorporated into the pact.

I was fascinated by Antonio's skill. His actions epitomized the kind of contradictory participation that the pact sought to incorporate. Under different circumstances, the refused handshake would have created deep political rivalry, perhaps even enmity, even Antonio's subsequent exclusion from public programs. It could even have unleashed gossip in town about a supposed armed confrontation between the two men to settle the issue. But the pact, and the consultant, could accommodate Antonio's refusal. Since it was not designed to distribute any resources, nor to promote a shared "solution" to water problems, Antonio could disrupt and find his own way of making the pact his own in front of friends and colleagues.

A couple of weeks later, I met with Antonio again. He picked me up on his motorcycle from the central bus station in Russas, the town next to Limoeiro. We rode to the local school, where he had arranged for us to use a classroom for an interview. Our conversation was engaging; it focused on the familiar points that people in Ceará use to explain the history of water: corruption, inequality, lack of rain. Once we were finished, he invited me to lunch at his house. We got on his motorcycle again and five minutes later arrived at his place. Antonio's wife came outside, welcomed me, and immediately asked if I would like to take a shower. It was about noon and the temperature had risen close to 40° Celsius (104° Fahrenheit). The high temperatures in Ceará's hinterlands make the offer to take a shower (*tomar banho*) a common one when people welcome you to their homes. After doing so, I sat with them at their dining table to have lunch. As we ate, Antonio told me more about his views on the pact. He was convinced that one could not give in. Participating in something like the pact without reservation was dangerous. Despite their aspirations to do things differently, projects like the pact inevitably embodied entrenched political hierarchies that had to be carefully managed. Favor trading, benefits for large landowners, those kinds of things, he noted, were always part of water politics, and one could easily, even if unwillingly, be enveloped by those webs. But one could not simply be absent from them either. That would mean leaving all the resources and possibilities in the hands of just a few people.

Antonio was describing a form of involvement that made explicit the doubts and reservations one harbored, instead of suppressing them. Putting one's doubts front and center gave the right impression, that one is a political actor with ideas of one's own, not a docile body that could be easily coopted into a collective controlled by somebody else. Antonio is also aware that this form of ambivalent engagement is not always welcome; sometimes it results in trouble or in being excluded. But it is a risk worth taking to develop a reputation as an independent thinker and political actor. In the end, Antonio thought that despite the pact's embeddedness in Ceará's troublesome water history, it had a peculiar spirit. The pact allowed him, and other participants, to assert their independence because it did not require them to erase their sense of being a political self, to ignore their ties to their municipality and the political relations they had there. "That is why I will work with the pact, because it really is not completely theirs," Antonio told me.

Ernesto wanted the rest of the team of consultants to think of the pact the way Antonio did. If technocrats and bureaucrats came up with a particular solution to water issues, people like Antonio were likely to point to how their solution allocated costs and benefits among members of their networks, and most likely would reject the proposal. On the contrary, if there were space for people like Antonio to bring his own suggestions, from the specificity of his political obligations in his town, his municipality, and the demands of his constituencies, something different could be possible.

At the next pact meeting in Fortaleza, after the handshake that never was, the French consultant recounted his experience with some bitterness. After decompressing with a couple of jokes and blushing once, the consultant discussed the Limoeiro events and Antonio's performance by explaining the political climate of the region. He explained how Antonio was caught in his own political web and how he had aptly extended the reach of that web to take hold of the meeting. Ernesto commented on how seasoned politicians know very well how to do that. But he then went on to note that if you are really committed to letting specificities flourish, these kinds of things need to happen. These are the kinds of differences that cannot be merely tolerated; they have to be celebrated. Ernesto gave his analysis of this nested political complexity as the air conditioner on the wall continuously blew air from behind, displacing the sheets of paper he had placed in front of him. Those of us seated around the table listened respectfully, and while some of the team members did not seem persuaded, they deferentially accepted his point. But if Antonio did not want to be fully incorporated, and Ernesto and his team were satisfied with his partial involvement, what was the pact really gathering?

In liberal democracies, the subjectivity of an individual—their sense of self and its recognition as a node of knowledge, skills, and social relations—is the taken-for-granted political unit that is then subsumed under a larger collective entity that we call community, society, nation, or state. A sense of belonging, or its lack, is analyzed by noting the extent to which the individual subject is added to equivalent units in order to establish social relations that result in different forms and degrees of reciprocities often coded as rights and obligations. Scholars have explored myriad concepts to capture this form of summation. Networks, for instance, are seen as transactional vectors that associate subjects and things around collective matters of concern (Latour 2005). Assemblages have been used to highlight the fleeting character of globalized interpersonal and institutional formations that hold the power and flexibility to redistribute political and economic capacities (Ong and Collier 2005). The multitude has helped conceptualize the unformed yet existentially powerful collective whose revolutionary potential is always on the verge of being actualized (Hardt and Negri 2005). And, of course, more classic notions such as the nation-state (Anderson 1991), community (Hayden 2003), and tribe describe the ways in which people come together to forge collective lives with others—human and not.

In the pact this kind of transition from individual subject to collective encounters a bump. In an unusual turn, the pact does not expect to include individuals as whole entities that can be encapsulated or summed up in a political program. The pact is after another form of collectivity, one that does not begin or end with the figure of the individual as a subject of rights. The pact centers on a smaller unit, the promise, a commitment that does not require a subject to belong, a member to be incorporated. An individual's promise is enough to be part of the pact; the pact does not demand much more. That distinction, between enrolling a subject and enrolling her promise, allows for a process that gathers without encompassing, for a form of sociality that lies somewhere between intimate attachment and unconcerned detachment. This form of collective allows Antonio to *take part in* the pact without *being a part* of it. Making sense of how one takes part, without being a part, could lead us into the metaphysics of subjectivity but I want to focus our attention elsewhere. As I learned from the pact team, this peculiar form of taking part is made possible by the material form of the promise and how that form enables its inscription, geographic movement, and aggregation. The next section focuses on that form.

COLORED SLIPS OF PAPER

One morning, Carolina, another pact consultant, met me at Avenida Abolição along Fortaleza's coastline with João, a veteran driver from the Assembléia. We were going on a five-day trip through the Mid-Jaguaribe watershed to conduct three pact meetings. As soon as they spotted me, and once they found a place to park, our first task was to find space for my bag in the bed of the pickup truck that would take us through more than one thousand kilometers of Ceará's highways. João jumped in the back of the truck to uncover a tight arrangement of a dozen medium-sized boxes that he swiftly reorganized, untying and retying them, to fit my bag. Half of the boxes were filled with booklets describing the pact process, which would be distributed among all participants as a way to inform them about the history and the methodology of the pact. The other half of the boxes contained hundreds of colored pieces of paper—light green, blue, yellow, and pink—that would become the material form of the promises.

During the previous week, the female office personnel of the pact had spent hours cutting sheets of paper to produce slips that looked like oversized Post-its. They then organized them into stacks, making sure there were equal proportions of each color. They proceeded to group markers and masking tape next to each stack and put each set in a separate box, one for each pact meeting the team was going to facilitate during the upcoming five-day trip. The thousands of pieces of paper the office staff had produced were going to be the infrastructure on which promises would proliferate and be aggregated.

Many hours later after Carolina and João picked me up, once we arrived at the location of our first meeting, we quickly spotted a place to hang the plastic banner that announced our presence: *Pacto das Aguas: compromisso socio-ambiental compartihlado* (Water Pact: shared socio-environmental commitment; see figure 4.1). João brought the box marked with the name of our location to the auditorium. Another team member that met us in

Figure 4.1. Banner announcing the pact-making event.

that location opened the box and placed the contents near the area reserved for the facilitator of the event. An hour later, after inaugural formalities that included salutations to every single authority present, Ernesto gave a rousing introduction to the pact that included his views on contradiction and the reference to Paulo Freire. Next, the facilitator took the floor and instructed the audience to divide into smaller working groups and explained the methodology we would follow. Carolina took advantage of a pause in the instructions to remind participants they should write only one pledge per piece of paper. That "keeps the ideas more mobile," she said. Carolina was anticipating the next step in the pact, the task of identifying promises that could be replicated across scales.

Each subgroup of five to nine participants received a stack of colored papers and three or four markers, along with detailed written instructions on how to proceed. People were told to reflect on the places they were coming from and to think about what they had to offer to improve the water situation in their communities, what they could do to help water reach those who needed it, what they could do to create a future without the *crises hídricas* (water crises) that filled their history. In response to this task, a school-

Figure 4.2. Organizing promises to conduct local pact meetings.

teacher spoke about the urgent need to make environmental education a required subject. She committed to lobby the regional director of her school district to make the classes she taught on her own initiative a required subject in the municipality. An agricultural extension officer promised to implement a training module to teach farmers about more efficient irrigation technologies. The sanitation secretary of the municipal government pledged to coordinate with his counterpart in the neighboring municipality to request federal funds to build a joint water treatment plan. A representative from a Catholic NGO promised to coordinate the actions of their water-tank provision project with the priorities set by the state-run community aqueduct program.

As promises started to flow within each group, they were written down by one of the women who, without any discussion, automatically became the group's secretary. At the end of the forty-five minute period, each group generated its own set of colored slips of paper. Promises and promise makers had been entangled through the speech act of the promise, but more importantly for the purposes of its endurance in time, through its inscription on two pieces of paper—one capturing the promise and another the name of the promise maker. The promises and the names of their authors were placed next to each other, stuck to a whiteboard, or more often to the walls of an auditorium, using dozens of small pieces of masking tape that the support staff conspicuously cut, turned into sticky loops, and placed on the edges of unused tables while the group was busy conceiving of their collective commitments. Once all the promises were up on the wall, looking at them gave you a sense of accomplishment, you saw a convergence of wills despite the radically different contents of the slips.

After the official meeting was over, the consultants or the ad hoc secretaries transcribed the promises on the slips of paper into Word documents while another team member took pictures to make sure no information was lost. Finally, the pieces of paper were taken off the walls, grouped to make sure each promise and the name of the promise maker followed each other, and put in a new stack that made it back to the box. Once in Fortaleza, the stacks of promises were stored and the digital files combined to produce a pact document, in this case for a particular municipality. This process was replicated dozens of times, for each municipality, watershed, and finally for the state as a whole.

Those colorful rectangular pieces of paper worked as Post-its do in analogous settings.[8] The popularity of these artifacts in development, design, and planning circles has exploded in recent years. Their trendiness is due, in part, to their capacity to carry and accumulate information in situations where there are too many stories at play, where too many ideas coexist. Due to that capacity, the media scholar Shannon Mattern (forthcoming) conceptualizes these kinds of singular units as "small intelligent moving parts" that have not only data attached to them, but also their cultural milieu.

The first appearance of these small intelligent parts in European settings has been traced to the transition from the nineteenth to the twentieth century when European thinkers entertained the idea of creating a knowledge management system that could handle large quantities of information without surrendering to the narrative limits of the monograph. One of the ways in which this utopia of infinite knowledge was organized was through the invention of the index-card filing system: "one work, one title; one title, one card," each "deal[ing] with a single intellectual element only" (Otlet 1918: 149; 1920: 186). Within this filing system, the index card had a dual capacity. It documented something, a piece of information. And, at the same time, it was a building block of something else, a constituting element of another entity (Mattern forthcoming).

As units, these small moving parts maintain their singularity. Given that capacity to remain singular, move, and build something bigger, we are better off thinking of slips of paper as more than decontextualizing organizational devices. They operate as "creativity machines" (Wilken 2010: 9) that do more than just reproduce what is inscribed on them; importantly, they do so without completely erasing their contents or burying them under the significance of a larger entity once they are grouped. In Ceará, each slip of paper is an active trace of connections, conflicts, debts, and enmities. It is as if these pieces of paper, despite their extremely brief contents, remind people that there is always more, that there are dense sociomaterial relations that the promise maker comes from and will never leave behind. In this sense, the promise/paper slip is not an abstraction; on the contrary, it is a dense inscription, a trace of entanglements that cannot be straightened out into a regulatable form because it is a record of all the social relations and histories that it represents but cannot contain. The people that participate in the pact never lose sight of that density, as we learned from Antonio.

From this point of view, the promise written on a slip of paper is a trace of social relations, an indicator of future possibilities, and a material object all at once. And while as anthropologists we often focus on the first two—trace and indicator—the unassuming materiality of the colored slip of paper is crucial for the possibility of its aggregation into a collective. In other words, the colored paper slips make possible the gathering of moral commitments to care into a multiscalar coexistence. The slip allows pact makers to aggregate without having to sum up into singular policies, projects, or laws. The material stacking of the promises, with their mobility and individuality, marks their capacity to be a constituting element of something larger while also remaining distinct units that cannot be merged or combined.

Promises, while always touching upon intimate fibers, can have multiple intensities. Promises between kin express an extensive "fabric of moral engagement, including the conflicting responsibilities and punishing demands" that the obligation to be available to others can impose upon us (García 2014: 52). At a less intimate level, the state makes promises for the future via plans and projects that are presented as promises that seem to always be "slightly out of reach, the[ir] ideal outcome always slightly elusive" (Abram and Weszkalnys 2013: 3).

The promises at the core of the pact are somewhere between these two. They are elusive because their future-oriented effects are impossible to fully verify either in the present or at every scale at once. But at the same time, each promise is intimate, a person-specific pledge, putting the emphasis on the body of the public servant rather than on the state as an abstract entity. It makes the intimacy of the state available to others.

AGGREGATING PROMISES

By 2009, two years after the idea was initially conceived, the pact organizers celebrated the fact that nearly every municipality, watershed, and the state of Ceará as a whole had conducted their pact-making activities. According to official records, thousands of people were involved in the process (Assembléia Legislativa do Estado do Ceará and Conselho de Altos Estudos e Assuntos Estratégicos 2009: 17). Through more than two hundred events, an unimaginable number of performances of political prestige gave shape to this attempt to break with history, make care for water universal, and transcend the elusiveness of state-centered, technical planning.

One of the pact's claims to power is precisely that it brought together the promises of an unprecedented number of people, more than eight thousand. Yet from my point of view, its potential rests not so much in the quantity of people involved per se but on how the promises of those eight thousand participants were aggregated. What makes the pact interesting is that its form has the capacity to make a promise "aggregatable" and, at the same time, allows it to remain faithful to its particular scale and specific context. Its social significance lies in its coming together as a gathering of place- and scale-specific social relations. Promises became collective as people followed a set of instructions, made verbal pledges, wrote them on slips of paper, came together to see them exhibited in walls, received electronic documents with their transcription, came together again for more meetings at different scales, refused to shake hands, reminded each other of the coffee they had together at the only local pact meeting they attended, or of the memorable joke that somebody told at another pact event. It is the gathering of all these that makes the form of the pact unique.

In everyday language, an aggregate is the gathering in one way or another of a group of entities (Potter 2004). The term evokes the notion of a herd or a flock, the act of adding a units to a group. Thus, aggregation brings units together without dissolving those constituting elements into the larger entity they constitute. The privileging of the unit in the aggregate is possible because aggregates have a certain transience to them—they seem to lack permanence. An aggregate can be undone. It is open for the inclusion or exclusion of units. It is a collective that is effective and precarious at once. Just as a flock of birds expands and contracts, dividing and merging, the aggregate never hides the fact that its constituting units are not attached to the group.

But despite their potential openness and light structure, aggregates have a negative resonance in the history of anthropology.[9] They are often taken as constructions that dismiss close social ties, homogenize contextual specificity, and erase difference. Yet in the twenty-first century of the internet, social media, and finance, aggregation has gained relevance as a native concept (Coddington 2015; Nafus and Anderson 2009) and political form (Juris 2012). Aggregates have returned as tools to let difference proliferate. They are more fluid than other formations we have used to understand collective life—such as community, family, and nation—but they do not replace those notions. Instead, aggregates exist alongside them, sidestepping the problem of contradictory allegiances and the difficulties of exclusively belonging to one group and not another, to one municipality and not another, to one watershed and not another. Aggregates allow for looser affiliations that can be transient, easily done and undone, plucking commonalities and counting them selectively for a specific temporary purpose.

If the methodological logic that guides the pact relies on the identification of promises that can be grouped and aggregated, it is not as a logic of accumulation that aims to add parts until they can complete a whole. The pact is not simply a summation of individuals into a new whole, it is a gathering of water-related promises. The difference between those two logics—the logic of accumulation as a summation of individuals and the logic of aggregation as a gathering of specific promises—can be visualized by comparing two images. One image is classic in political theory, the other is closer to algorithmic and mathematic forms of aggregation.

The first image relies on clear units and addition to achieve unity. We find this peculiar sense in a popular illustration of Thomas Hobbes's *Leviathan, or The Matter, Forme, and Power of a Commonwealth Ecclesiasticall and Civill* (1651 [1991]; see figure 4.3). In this illustration we can see how, at the end of the day, the collective is an all-encompassing whole where the individual is assumed to be a smaller part of a larger unified entity represented

Figure 4.3. Detail of *Leviathan* illustration (Hobbes 1651 [1991]).

by the ruler. Leviathan is composed of the sum of clear units, the whole bodies of its subjects. Those subjects are clearly incorporated, distinctively conceptualized as belonging to the body of the ruler. For Leviathan, its constitutive elements are, on the one hand, self-evident, and on the other hand, clearly internalized. Here, the body of the subject is a familiar singularity that belongs, in all of her facets and despite any intimate misgivings, under the larger body of the ruler. Many of our analyses of collective projects in liberal, capitalist, and settler societies rely on this image. But we know that this idea of membership relies on an idealized sense of belonging that is taken for granted and that fails to grasp people's multiple and sometimes incompatible allegiances (Simpson 2014).

I see in the pact another way of posing the question of collectivity and belonging. If we were to visualize the pact, it would be through the form of an aggregate that does not depend on the incorporation of self-evident units. Andy Lomas, a digital artist working in the United Kingdom, has produced such an image. Unlike Leviathan's illustration, Lomas's image does not offer us an easy depiction of its constitutive units (see figure 4.4). Are these trees? Leaves? Bacteria? In his image it is difficult to elucidate any uniform units; we are forced to take a step back and ask what the units are and what quality of property is bringing them together into a set. Here, to be thinkable, aggregation necessitates the identification of the parameters

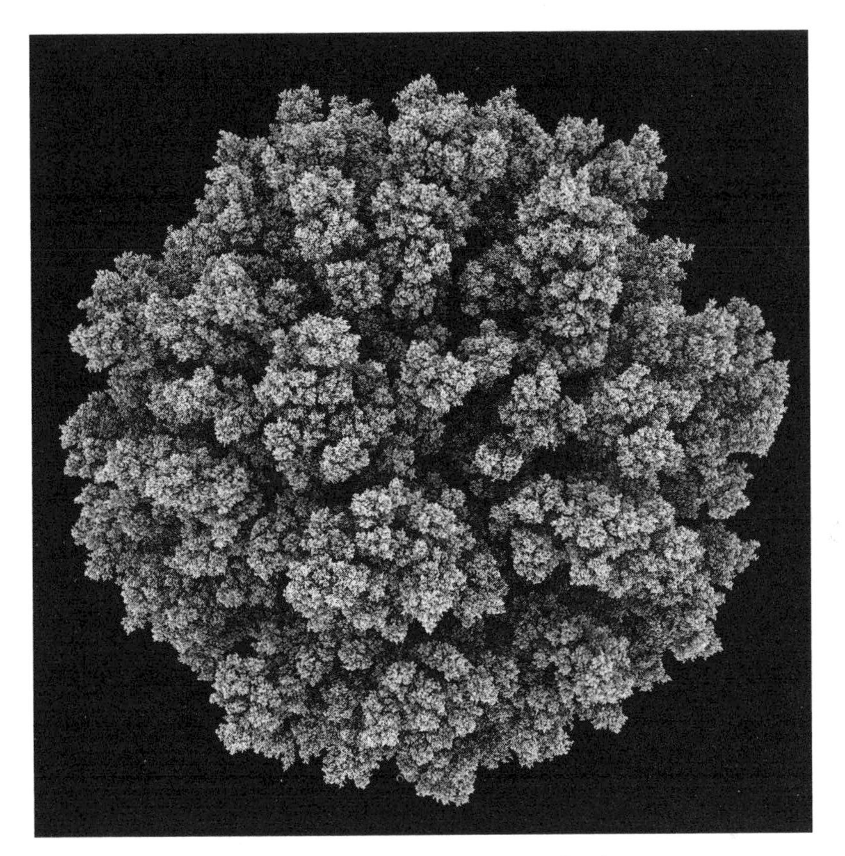

Figure 4.4. Image of a mathematical aggregate. Courtesy of Andy Lomas.

that make the gathering possible. We cannot automatically see the summation of clear self-evident subjects; we need to ask ourselves by what method the elements were selected and gathered.

What Lomas's art makes apparent is how important the criteria for the aggregation of a set of elements is, the foundation of a gathering that is only temporary and not naturalizable. Lomas creates in the observer the need to know how a group is brought together, as the aggregate only makes sense if we know what makes something count as a unit, despite our uncertainty about what exactly those units are. The moment our attention drifts away from the selection criteria, the possibility of perceiving the aggregate disappears in front of our very eyes. It is no longer recognizable as a collective, it becomes an inaccessible form.

Ceará's Water Pact resembles Lomas's image more than that of Hobbes's

Leviathan. For its organizers, its power to bring about a society that cares for water and comes closer to universal access lies in its selectivity, in the capacity of its method to collect and activate a peculiar unit, a commitment to care for water, without aspiring for participants to identify holistically with the pact, much less requiring any form of membership from them. If we lose track of those methodological matters, the pact becomes invisible as an event that matters.

As we have seen, the promises, the basic units in the pact, cover all sorts of issues. Additionally, talking about a single pact is misleading because there were multiple pacts at the municipal, watershed, and state level. The only clue we have to figure out the form that the aggregation of promises takes is their initial inscription on a colored slip of paper. Beyond that, the promises have almost nothing in common. Those people, relations, and materials gathered through promise-making rituals disperse into their own contexts, political webs, institutional feuds, personal likes and dislikes. The pact, as a gathering of promises, is at once transient and permanent, recognizable only as long as we keep in mind the form of the promise.

CONCLUSION

Practically speaking, the pact asked for people's commitments to solve the imminent problems posed by water scarcity, pollution, and climate change. It proposed promoting people's care for water as a way to leave the history of politically commodified water behind. Care, inscribed in the form of promises, would move Ceará closer to a humanitarian redefinition of water management, and possibly of infrastructure and of all those social relations undergirded by water. Methodologically, the pact consisted of a series of public rituals where participants publicly made pledges. And, materially, the pact assembled prodigious quantities of written records, later exhibited as published documents and PowerPoint presentations. The organizers of the pact launched the initiative with the expectation that they could break with history. Originally, they hoped the pact would differentiate historical forms of commodification from humanitarian and universal commitments to care. And yet, in a sense, their efforts brought about a form of obligation that looked very similar to the history that they wanted to break with.

Another of the pact's peculiarities is the way in which it searches for the "social." Usually, state-led initiatives search for the social by going to society to find churches, community organizations, businesses, and everyday citizens. In a break with this paradigm, the pact organizers looked for the social within the state itself—in the capacity to care of public officials, in the promises they would make to each other. It is as if in their effort to rechart water as a social substance, instead of seeing society as the accumulation of citizens with a particular socioeconomic status, ethnicity, and kin relations, they visualized society as a collective of promises, of publicly declared responsibilities, that state officials at different scales were responsible for. To put it succinctly, the pact was a device to change society and create the future conditions of water by gathering pledges made by public servants. Contrary to the original plan, society and the state became indistinguishable from each other.

By staying with the form of the pact and thinking about the possibilities its organizers see in it, we discovered a gathering in contradiction. Rather than being one thing, the pact organizers intentionally wanted it to be many things at once. If the promises people made across different scales did not fit together and were in contradiction, that would strengthen any chance the pact had of setting the preconditions of the future. Thus, context specificity and the contradictions derived from it were precisely the traits that allowed the pact to be a collective that is not a summation of individuals. Rather than being a mechanism for transforming differences into commonalities, as the concept of universal rights does, the pact was an arrangement to generate contradictory specificities.

Years after the culmination of the pact-making process, people remained unclear about how to assess its accomplishments. The latest report the council produced (in 2013) noted that only some municipalities, around forty, continued to use the pact to organize their actions. In an effort to show the pact's significance, the report combined information on specific projects launched at the municipal level, percentages of completion of state water initiatives, and indicators of the actions taken by regional agencies. The names of the promise makers, however, had disappeared. Agencies, secretariats, and city governments became the promise makers.

Since that first experiment, Ceará has organized two more pacts, one focused on drugs and violence and another to organize the strategic planning of a new port in Fortaleza, Ceará's capital city. Further, in 2013 Brazil's federal government launched a country-wide pact for water management, and later a new NGO in Amazonia was created under the same name *Pacto das Aguas*. So, by the end of the second decade of the twenty-first century, the Pact as a social experiment was proliferating.

In 2014, I met Rodrigo for lunch in downtown Fortaleza. Rodrigo was one of the pact consultants and a close collaborator of Rebecca's. He often facilitated local meetings and went to do the same for the two subsequent pacts Ernesto organized. The main topic of our conversation during lunch was the pact. We were both nostalgic about the energy of those days. Rodrigo felt the new pacts lacked the enthusiasm and participation the Water Pact achieved. I was curious about what he thought had been the impact of the whole process. When I asked him directly, his response was unsettling. He first mentioned a few specific projects that had been implemented. Then he quickly went in a different direction. Rodrigo reflected on the scale of the pact as an effort to break with history and change society. Clearly, there had been many difficulties in monitoring what the pact accomplished. But after a while he had realized that "maybe the magic of the pact was precisely that you would never be able to assess its effects because it was so all-encompassing and diverse. It was everywhere and nowhere at the same time."

At that moment I pondered the implications of his comments. On the one hand, I could hear Pedro and José's voices predicting that the pact was just a replication of what already existed. Another invention, another attempt to modernize water that "society" had not asked for. On the other hand, I remembered the hundreds of handshakes, conversations, promises, and displays that were brought into existence with the involvement of more than eight thousand people. Pedro and José were assessing the pact on the basis of its recognizable effects, actions that could be associated with recognizable causes. Rodrigo was reminding me of something else: he was pointing to the importance of that which is not recordable, of taking the risk of attempting to break with history, without being able to verify the effects of one's actions. Maybe breaking with history means precisely not producing another history to replace the original one.[10] The pact created the possibility of aggregating commitments to care, but that very task was impossible to verify. If all of society cared, such care could not be found in any one singular place. Rodrigo's reflection suggested that despite all the energy, effort, and hope that went into the pact, it remained a massive effort that may never achieve the full status of having occurred, and that was not a problem. Ultimately, whether the pact contributes to making water universally accessible can only be recognized at some point in the future. At that time, a history of water might recognize as significant what in a previous present was barely effective.

CONCLUSION

As I began writing this book, *National Geographic* magazine published a special issue whose cover consisted of a gray background with blue letters that read "Water: Our Thirsty World." The image was overlaid with drops that resembled raindrops over glass, transparent yet clearly delineated ovals. The issue was launched with an accompanying exhibit hosted by the Annenberg Space for Photography in Los Angeles. The array of topics covered by the pictures were representative of what has become a widespread understanding of the challenges that water poses to human technologies, the dramatic effects of living without access to clean water sources—especially for women—the ritual uses of water, and the apocalyptic predictions of the disasters that climate change and the loss of aquatic biodiversity pose to the world.

I visited the exhibit during its final days. The pictures that appeared in the magazine had been enlarged and hung on the outer walls that surrounded an inner circular screening room that showed interviews with the photographers and provided more information on the global water situation. The room was overflowing with people, nearly elbow to elbow. It was easy to see how taken people were by the beautiful images and the background music that accompanied them. After about twenty minutes of this media immersion, the lights were turned on and we remembered that we were at a photography exhibit and started circulating to see the large prints.

I gathered from the attendees' comments that the most powerful pictures were those that showed the "ritual" or "spiritual" uses of water and the ones that depicted the daily struggles of "African women" to collect drinking water. Picture the first series of images, titled "Sacred Waters." The *cenotes* of Mexico's Yucatán Peninsula that the Maya believed were pathways into the underworld; a baby being baptized in a Greek Orthodox Christian Church in Istanbul, Turkey; a parishioner cleaning a cross carved out of a layer of ice in a frozen lake in Maine; a Shinto man in Mie, Japan, standing under a waterfall surrounded by rocks and candles while a number of people observed his communion with water and the creative force of life (according to *National Geographic*).[1] Now imagine the second series—this one elicited even stronger reactions—titled "The Burden of Thirst." A group of eleven women, photographed from behind, as they walked into a barren orange and brown horizon carrying five-gallon plastic containers tied with ropes on their backs; a close-up of a woman's face standing inside a well and passing a bucket filled with water to a pair of hands lifting it to the surface; a black plastic water tank from which women and children filled former gasoline containers transformed into buckets so that they could carry them on their backs to their villages.

The images were moving. Their emotions were readily available to the audience. People left the exhibit both inspired and troubled by the global water situation. I walked out of the gallery also affected by *National Geographic*'s pictures (critiques of their orientalizing and objectifying tendencies aside). I felt, I must confess, inspired by the power of water to tap into people's affects and rationalities, and at the same time conflicted about my decision not to resort to similar images to explore how our worlds shape and are shaped by water. I had decided early on that my ethnographic account was going to focus on the dimensions of water that would never make it into *National Geographic* pictures. I would focus on the enormous amount of "deskwork" required to shape the worlds that are later photographed for *National Geographic* magazine. The exhibit served as both inspiration and disappointment. I found myself wondering how a *National Geographic* exhibit would look if it also paid attention to regulations, mathematical calculations, and collective political experiments.

This ethnography is an attempt to restage that deskwork and cubicle-based decision making that does not make it into exhibitions. It is also an effort to mirror the temporality of the political and technical worlds that bureaucrats, economists, lawyers, community representatives, and NGO managers live in. These groups of people are invested in transforming the conditions that give rise to the situations that *National Geographic* photographs, which are structured by assumptions of liberal rights and the dominance of market logics in the distribution and use of water. I trace their efforts to link water governance with a notion of the common good that is constituted by ideals of making collective decisions, acquiring new knowledge, respecting human rights, and keeping prices outside of markets. It is fair to say that my colleagues working on water issues agree with the statement that "one resists co-optation not by distancing oneself from power, but through the vigilant practice of not being co-opted" (Gibson-Graham 2006: xxxi). This book takes you to some of the instances that constitute that vigilant practice and opens up the technical devices that shape their everyday routines.

Many scholars argue that we need to attend to water via its uniqueness. That claim to uniqueness often justifies inaction in the face of the overly complex challenge of making its use, distribution, and treatment more sustainable and democratic for all forms of life, human and not. That uniqueness is also good material for *National Geographic* special issues. It generates something of a watery mystique, inspiring awe at the sublime overflows of its affective and material meanings. There is something else, though; the everyday and, in comparison, fairly unexciting matter of the technical concepts that organize water's availability. I have intentionally focused on these concepts in this book. I have set out to show how water is far from being unique and is thoroughly embedded in long traditions of law, economy, religion, and liberal politics. Thus, in a way, I want to argue against the exceptionalism of water, a condition that can effect a mystification that closes off the quotidian forms in which its materiality is affected. I want to remind us that the politics of water are broad, transversal, and not exceptional. And as we recognize that excess, we need to remember that there is more available to us than the power, profit, poverty paradigm (Weston 2016) and the mystified regime of uniqueness. The middle ground is full of fascinating configurations.

In 2016, 2.1 billion people in the world lack access to clean water, floods come to unexpected places, and pollution turns what for some are quotidian acts, like taking a shower or brushing your teeth, into acts of toxic exposure. In the not so distant future, the transubstantiation of water from ice into a warmer, hence expanded, substance will lift the planet's temperature and redraw the contours of oceans, lakes, rivers, and land. Water is a romantic substance that reminds those humans who have forgotten it that their cells are as material as those of rocks. It is also a commodified substance whose value perplexes the observer of capitalist endeavors. It is a scientific riddle, with molecular properties unlike other substances. It is a substance that pushes all sorts of boundaries at the conceptual, pragmatic, embodied, and affective levels.

People in communities all around the world face water problems that are as great and unwieldy as they are unequally experienced. Political actors, celebrities, activists, artists, CEOs, children, and elders are moved by the effects that water lack, excess, and contamination have on the everyday lives of millions of people in Latin America and elsewhere. Not surprisingly, we hear continuous calls to do things differently. From Pope Francis to Berta Cáceres, an indigenous Honduran ecofeminist activist murdered in 2016, we hear demands for renewed moral schemas to guide society's dealings with water. Those appeals for ethical renewal to restore the health of the planet ask us to empathize with a world that is much more than human. The biosphere, the geosphere, and all forms of life made possible by water are main characters in those calls for moral action.

In parallel to those calls for new moral schemas, we find philanthropic campaigns, movie stars, and even James Bond (in the film *Quantum of Solace*) concerned with the threat of a global shortage of clean, healthy water. These campaigns invite us to purchase goods and donate money to fund interventions in Asia, Africa, and Latin America. Buy this bottled water and one village in Africa will get a well. These initiatives are designed to extend already existing economic and environmental imaginaries through green and philanthropic capitalisms. They promise to address global inequalities by creating more commodities and larger markets.

In between these calls to radically renew our moral schemas and the invitations to extend modern capitalism to produce technical fixes, we find my interlocutors, people who strive for transformation without either jettisoning what is or keeping it all in place. They are individuals who consider ways of changing the future from within the technicality of already existing worlds and by using already available tools. They work to differentiate the world that already is from the world that should be. Their efforts articulate economic, legal, and natural histories to establish new responsibilities for a world and a global population that is now recognized as fragile and at risk of destruction. They are a group of actors with technolegal skills to manage water locally, while remaining cognizant of the planetary dimensions of the problems they deal with. They are those for whom critique is a performative act that does not consist of diagnosis but of world-making. They encounter the world and they act to modify it, if only in small ways.

By intimately following their work, this book has sought to answer three straightforward questions: How do people create distinctions in the worlds they inhabit? How do they navigate seemingly contradictory categories? And how are those distinctions and categories connected to their aspirations for the future? I have focused on the material-semiotic life of water, but I have proposed attending to it in spaces other than conventional water bodies such as rivers, lakes, oceans, aquifers, pipes, and reservoirs. Instead, I have taken you to bureaucratic offices, congressional discussions, and public workshops where work to redraw legal, economic, and political distinctions is constantly unfolding.

Throughout the chapters in this book I have shown how the work of creating separations requires a form of labor without end. As soon as a distinction is set in place to separate, for example, water as a commodity from water as a human right, the world challenges that distinction by revealing how what seems a bifurcation has only been a way to take the inseparability of those figures in new directions. And yet, for my collaborators this continuous slippage is not understood as a problem waiting to be resolved; it is the very shape of the world. They are used to having the distinctions they create reappear as new fusions, requiring in turn new dissections. That slippery nature as a fact of life is part of their explanation of how rights and commodities, those fundamental figures of liberal capitalism, seep through and endure despite mobilizations against them. This book has taken you into a variety of moments where, despite their temporary character, people attempt to instill those separations.

The things we learn by following their work are not limited to water issues. The task of dealing with contradictory classifications, values, categories, and desires is distributed among all kinds of collective endeavors. Struggles that juxtapose economic reasoning with legal arguments, the realms of rights and the realms of commerce, are spaces where people find themselves creating separations, clarifying what counts as one category and what counts as the other. These struggles are unfolding all around us, at the level of state activity and bureaucratic organizations, but also at the level of personal interactions, as when people consider how their actions affect the beings they care for. In Latin America, some of the issues where this struggle plays out include the social status of health care, minerals, access to the internet, food, information, land, computer literacy, and education. These concerns are dealt with collectively. Governments, NGOs, citizen associations, corporations, and many other social actors of late capitalism relate to these concerns and use them to create ethical distinctions that qualify the liberal heritage and its capitalist companion. In their everyday lives, people use these ethical separations to help clarify how their intimate lives and professional aspirations coexist.

The other question this book speaks to is how the work of living with proliferating fusions and bifurcations relates to the future. We have a habit of thinking of the future as a particular configuration that, even if not fully predictable in every detail, is somewhat imaginable. My interlocutors engage with the future differently. They do not pause to specify the details of the future they are creating, and yet they still have a clear sense that they are contributing to its creation. In other words, their view of the future is not predictive—neither in the speculative form of science fiction, nor in the empiricist form of science. In place of images of the future, my collaborators hope to create preconditions, forms of structuring future collective responsibility even if that responsibility cannot be allocated juridically, that is, by identifying responsible bodies and establishing causal connections. This nonpredictive relation to the future cannot be reduced to a vision of how things should be, largely because my collaborators' recognition that the differences they create can seem fairly ineffective actions, activities that those interested in documenting social change can easily discount as insignificant.

Such a reading of futurity and lack of efficacy needs to be qualified. The reason the bifurcations my interlocutors work hard to produce might not be recognizable as effective is that we can only read them from the visibilities of the present, from what we know and desire today. They do not necessarily contribute to our desired vision of how the future should look. They do not erase commodification, they do not purify the humanitarianism behind a universal right. That does not mean, however, that the conditions my informants and their devices help set in place will remain meaningless for a future that has not yet arrived. At that moment, it might turn out that their actions were powerful precursors, activated as new and significant elements of that present. Just as, for example, the statistician that produced the first consumer price index in the nineteenth century did not foresee how inflation would help determine what is a humanitarian price for water, my interlocutors do not claim to secure the consequences of their actions in the future. In this particular philosophy of history, people know that their efforts can be flawed or never fully effective, and yet are necessary. They also know that in the future, the significance of their acts will be recognized differently, not as an instantiation of an image but as one of the preconditions for an unseen one. For that reason, inaction in the present is not an option, even if for some observers what my collaborators do is nothing but that.

This way of creating differences and relating to the future necessitates a way of thinking about water that does not take for granted its materiality or the political valence of particular water forms, as if one could determine once and for all that free-flowing rivers are inherently good, and bottled water is inherently bad. I have attempted to show the vast worlds that are left unexamined when we assume that the politics of water are more "intimate" near H_2O, rivers, pipes, or reservoirs. Further, I want to suggest that we need an enhanced understanding of the materiality of water that includes things like colored slips of paper, goods purchased by statistically idealized households, speculative taxonomies of water bodies, and paper water bills. Attending to that extended materiality reveals how water can never be separated from the ideal forms that we use to describe it and prevents us from extending into the twenty-first century a naturalism that takes for granted the materiality of the world as a stable object.

What I have tried to show is how futures are constantly being produced out of mundane actions—for example, separating and dividing, reseparating and redividing that which seems entangled and ethically feels it shouldn't be—and how those actions are crucial sites where big moral and even philosophical questions are encountered and entertained by everyday people in Costa Rica, Brazil, and elsewhere. But these are not exclusive concerns of technocrats or activists. To an extent the everyday work of future creation via the mundane devices through which we live our lives is the everyday task of engaging the world and trying to elucidate what a livable life looks like.

The arguments I make about the politics of the future, the making of bifurcations, and the life of technopolitical devices are significant well beyond water. Formulas, indices, lists, and pacts, the tools through which my interlocutors relate to the future and effect differences, proliferate in all domains of collective life and in all sorts of environmental and political settings. They shape all sorts of bureaucratic and political spheres. I contend that it does not matter if one looks into the production of pleasure, the exacerbation of economic asymmetries, the management and imprisonment of bodies, or the invention of new objects, just to pick some examples.[2] These domains, and many more, are filled with devices like the ones I have focused on. Thus, my interest has been to explore what we can learn if we begin our analyses by looking at these kinds of objects rather than by affirming well-established categories such as subjects, relations, structures, or histories. Of course, my point is not that we should abandon those cherished categories but that the experiment of starting our analyses elsewhere offers new insights about their theoretical purchase.

Beginning my analytic work from the device as an alternate location led me to revisit our affinities for the concept of entanglements. I asked: What happens after entanglements are diagnosed? If, as feminist scholars have shown us, entanglements are the conditions of life and death, how do people live their lives amid generative and destructive embroilments? While we have paid attention to how those entanglements affect the world, we have not focused to the same degree on the possibility of living through disentanglements. Thus, I have argued that many ethical and political projects around the world are launched to separate things and keep them distinguishable. This is the case for the relation between a human right and a commodity. Such an analytic project is not reducible to what ideas of purification (Latour 1993) have already done in the social sciences. My objective has been to highlight the many forms that the everyday work of creating bifurcations and disentangling things takes. This, it seems to me, is a crucial way to enrich our understandings of how people live their lives amid profound and contradictory environmental, humanitarian, and economic demands.

This book is also an ethnographic experiment designed to create a particular analytic style. To achieve this style, I made a series of deliberate choices regarding the selection of the group of people and the devices that structure the narrative. I looked for entry points that allowed me to sidestep any opposition, implicit or explicit, between the phenomenological and the sociological, or between the intimate and the structural. In order to escape those inherited dichotomies, I looked for generative junctures where the opposition dissolved. I searched for the everyday actions where people openly attempted to transform structures with full consciousness of how their gestures, utterances, and immediate environments allowed them to do so.

That is how I ended up working with middle-range public officials, NGO leaders, citizen organizations, and politicians. These are people whose everyday responsibilities are not centered on how they might change their own lives or material conditions, even though of course those are at stake in their daily jobs. Rather, their quotidian activities consist of changing the conditions that shape other people's worlds, and only incidentally their own. They make, for example, administrative decisions that affect all citizens served by public utilities, or they craft laws that people then challenge or implement as they confront the limits of private property, or they convene subjects to be present in the making of collectives where political prestige and hierarchy are reproduced and possibly challenged. Beyond office politics and their own intimate senses of purpose, their responsibility is to Others, human and not, and that responsibility is not expressed in spectacular mobilizations. It is experienced in somewhat uneventful and boring everyday work.[3] This dimension of their engagement with water and with collective concerns is often glossed over in our analysis of political hierarchies and organizations, and for that reason I wanted to focus on their labor.

And yet my purpose in the book has not been to stabilize them as a new kind of "community" not yet described by anthropologists. I did not want to turn them into archetypes or exemplars of a type of subject. That would have led to a simultaneous flattening of their existence and of our imagination. What interested me was the means by which they connect, or not, everyday tasks to a sense of a future good, even if that sense is not an image but a set of technical responsibilities for setting the conditions of the worlds that Others inhabit (Robbins 2013: 457). During my research, those responsibilities and aspirations took the form of a specific set of devices.

DEVICES

A device is a kind of instrument that is highly effective in organizing and channeling technopolitical work. It is an artifact whose leverage to change collective worlds resides in its capacity to merge practices and desires with long-standing assumptions about sociality that have been embedded in legal, economic, and other technical vocabularies and institutions. A device is a structured space for technical improvisation; its seeming fixity requires people to imagine tweaks and hacks to take advantage of the play inherent in structure. As I showed through the chapters, a device is embodied in the actions of specific persons, but it is also a braiding of long histories. A device affirms and destabilizes social categories and institutions, while allowing us to identify the particular practices, offices, computer files, and conversations whereby that material-semiotic labor is performed. I deem the device an important political participant in social life that has received little anthropological attention.

After studying these devices carefully, I came to understand their power as twofold. First, a device has the capacity to unleash effects in the world. Once activated it may, for example, channel the energies that shape regimes of value, help argue judicial decisions, create associations with material objects in the world, or become a vehicle for creating new obligations. The concrete material effects of a device can be sociologically traced through many settings. Second, a device is also a conceptual arrangement of contemporary and historical ideas into material forms. It is a lump of concepts that have been tied together through specific practices that give them an identifiable presence. A device carries with it the epistemic histories of an ontology, even if those histories are not explicitly invoked as its predecessors.

As people shared with me how they went about dealing with the responsibility of effecting a difference in the world, it was remarkable how their descriptions did not consist of image-like renderings of utopian or dystopian environments or societies. Rather than painting such images, they centered on the specific devices they were using. They pulled me deep into technicalities, reflecting on the adaptations and changes they were making to their tools. In one well-entrenched form of reading their work and lives, we would interpret that focus on their devices as a way to escape the fact that they work in public offices that never fully solve the problems they were originally created to address. And while we might be tempted to reproduce that story, I realized that such a story falls short of the ways in which their actions touch upon the world. Thus, rather than writing a diagnostic story of shortcomings, I explored what was at stake in that disciplined focus on their devices. The result has been that slowly the device became more than my ethnographic entry point. It became my analytic framework. This choice was also a way to stake a theoretical claim within an anthropological tradition committed to beginning any theorization out of one's informants' analytics rather than taking their everyday lives as objects waiting to be diagnosed and interpreted with imported conceptual resources.

The product of borrowing their thinking to create the analytic structure of the book is this collection of four devices, three from Costa Rica—formula, index, list; and one from Brazil—pact. The Costa Rican devices are all highly technical instruments, hidden behind complicated processes that I followed for a long time in order not only to understand what was going on but also to grasp their ethical and political intensities at different moments. It is not a surprise that the life of those devices goes on in the back corners of bureaucratic offices and in Kafkaesque congressional procedures. Costa Rica's environmental politics, and really almost all mobilizations to address collective life in relation to the state, are today experienced in this way. There is little or no sense that all-encompassing change is possible. Rather, there is a sense of being "stuck." If, by a stroke of luck, transformations are brought about at a collective scale of political action, those transformations are piecemeal, only one tiny step at a time.

As my research in Brazil progressed, it was striking how different the Ceará experience was from Costa Rica's. While, of course, similar technopolitical devices like the ones I encountered in Costa Rica are at work in Ceará, the circle of water activists and experts among whom I worked were all touched, more or less intensely, by one process: the Water Pact. It was notable how the pact was choreographed around the possibility of all-encompassing transformation. The pact retained as a viable possibility massive change, even if that change was far from certain. The philosophies, histories, and techniques pact organizers relied upon to bring it into existence were marked by what they understood as "large-scale" visions to aggregate the political will of "all of society." This device affirms a notion of totality that is still transformable as a collective, although not changeable in the modernist understanding of centralized planning that for long was characteristic of Brazil. Today, all of this is at stake as the country grapples with the rise of extreme right ideologies, homophobia, and misogyny as a valid political platform.

I curated four devices with such different political tones and scalar forms into a group to perform a precise type of critical anthropological exposition. I wanted to retain the sense of fragmentation and lack of closure with which my interlocutors experience their worlds. That decision turned this monograph into a "performative scholarly engagement the enactment of which constitutes its critical currency" (Maurer 2005: 25). This form of critical analysis intervenes in the world by performing that which it argues. In other words, rather than diagnosing shortcomings, it attempts to restate the methods and practices by which such shortcomings can be countered, no guarantees offered. By mirroring the very form of political labor my interlocutors put into shaping the material semiosis of water, the book extends an invitation to consider their worlds from within. This aligns me, possibly in an awkward way, with the project of immanent critique, but also, and perhaps most significantly, with a fundamental anthropological conceit: the idea that as a scholarly field and as an existential form, anthropology can approximate and inhabit worlds without explaining them away, without reducing them to already existing schemas in ways that flatten their irreducible uniqueness (even if those worlds are technocratic!).

One of the implications of committing to a performative analytic is that aesthetic elements do all sorts of theorizations in the background, by virtue of their mere existence. One of those "background theorizations" embedded in the form of this book is something I think of as a magnification effect. Such magnification enlarges something that we usually perceive as of a smaller size. This form of augmentation is different from moving up through supposedly self-evident scales, such as moving a description from a particular event to a historical trajectory, or from an individual experience to the story of a community. The change in scale I refer to here by using the word magnification is tactical and for the purpose of analysis only. It is the effect one accomplishes when putting something under a magnifying glass to gain a different insight into its details, and hence encounter it differently. This is a distinct form of focusing our attention.

I have magnified each device to the extent that it might seem that I am overemphasizing its power, rendering each as occupying a greater slice of social life than it actually does. And yet, as we make this evaluation comparing a supposedly aggrandized social significance to a more proportional scale that is more adequate to the "real," one question remains: What parameter determines what is its "adequate" size? Put differently, we can only see aggrandizing as problematic if there is another scale that we accept as adequate. But what is that other more adequate scale? Is it the social system? The phenomenological experience? An institution? A ritual? By magnifying these four devices, I hope to create some instability in our familiarity with particular proportions and scales of analysis.

This magnification effect can also lead us to assume that these devices have great causal powers on their own. That is not the case; they are one juncture made possible by many other instruments, hegemonies, logics, and affects shaping water worlds in Costa Rica and Brazil, and for that reason I do not claim that these devices hold any monopoly over causality. What I have highlighted is their capacity to mobilize and direct histories, desires, materials, and concepts, and in that functional multiplicity affect the world. I am convinced that potential needs to be studied by anthropologists. These devices cannot be the exclusive realm of the professionals who design them—economists, lawyers, political scientists, business managers, and so on. To make them ours—that is, to turn them into anthropological objects—we need to magnify them and in the process recalibrate our assumptions about what constitutes adequate scale, domain of analysis, and even causal relations. Too many forms of life are affected by these kinds of devices for us to not be curious about their workings.

CRISIS, APOCALYPSE, WONDER

In the twenty-first century, the political valence of water is shaped in humble and spectacular ways by humanitarian and commodity logics expressed through a narrative of crisis and apocalyptic end of times. It is tempting at this point in history to frame the conversation by drawing out those undertones in our analyses of water issues that are so asymmetrically experienced.

I have tried diligently to avoid the temptation of using that narrative frame. In the introduction I explained the reasons why the crisis language, with its particular understanding of history as a sequence of pivot points, was inadequate to understand my interlocutors' work. Here I refer to why the underlying sense of apocalyptic end of times due to climate change is also unsatisfying. Apocalyptic thinking highlights temporal disorientation, but not in relation to a lost future as one might think. Instead, it signals a "hyperbolic anxiety that the future may now be unattainable because the present fails to bring the past to utopic completion" (Wiegman 2000: 809). The apocalyptic instills a sense of end of times that depends on the existence of a previous definition of what is or should have been in the future. It depends on an implicit certainty about the existence of some vision from the past that has ended, that will not become.

That apocalyptic disappointment is only possible if you knew what the future would look like. But the problem is that such certainty about what the future should have been is available to a very small group of people in this world. It is a very specific cultural formation, even a subject position, that cannot be assumed to be universal in any sense. What I learned as I moved further and further away from apocalyptic crises is that the practice of lingering over diagnoses of catastrophe and the end of times is mostly available to distant observers. People having to come to terms with dire circumstances, those that we might label anthropocenic doom, extinction, utter dispossession, or violence, seldom see their histories as already determined, as interrupted becomings that have been settled.

As an anthropologist from Latin America working in the United States, I experienced a particular attachment to the worlds I was part of during my research. That attachment took the form of a reluctance to close off the narrative. After doing some reflection on the theoretical foundations of this attachment, I realized that it was impossible to think of these stories as events that I went and observed "in the field" and left behind as I returned to my position at a university in the United States. In a way, this was so because the stories that captivated my imagination felt personal to me, not by choice—that is, not because I developed close ties with my interlocutors (which I did). Even if I had not developed those relations, those stories and the devices at their center would still feel very close. To bring the point home more clearly, think about the image of a water bill that you saw in chapter 1. That bill is my mother's water bill for a random month in 2011, before AyA shifted to electronic billing. The answer to the question of how much she pays for water and how that relates to humanitarian aspirations is still a work in progress, answered cyclically, and each time requiring the work of elucidation. In this very concrete way, the devices I have analyzed, and the worlds they are part of, are themselves stories that I am aware continue to unfold; they are worldly narratives without a known end point, hence my reticence to close them off.

Thus, even if I exerted my full will and tried to imagine myself as a subject of another political and environmental milieu, which in a sense I also am, I would be unable to effect a closing. I have written this book from that epistemic position; I have tried to answer questions that I hope speak to the reader but that are motivated by the kinds of questions that fellow activists and bureaucrats themselves ponder as they go about the difficult job of keeping legal fictions and myths of liberal collectivities as realities that are agreed upon and contested at the same time.

Beyond the difference that personal life history makes, there is something else behind my decision to commit to non-finalized stories. For me, rejecting closure is also a way to establish one's relation to the empirical. As I said, the devices I analyzed continue to lead social lives. Activists continue to invoke the list Libertarians produced; prices continue being adjusted using the inflation rate; the ethics of profits are still subject to numeric elucidation; and promises continue to bind people to collectives that are tenuous. Thus, what could have started as a personal attachment due to a particular life history turns into a mode of theorizing how anthropologists participate in unfolding events. Closing off histories that I know continue to move became empirically inaccurate and politically fraught.[4]

In order to embrace this rich technopolitical universe without flattening its uniqueness or closing off my collaborators' futures, I invited the reader to approach the book the way I approached my research, and ultimately my writing: with the openness of wonder. Rather than a topic, wonder was the epistemic mood from which I set out to understand these devices and write this book.

The sense of wonder I called for is far from any form of miraculous devotion or positive admiration. Rather, the sense of wonder that I cultivated was the one we experience when we look at the world around us and see the strangeness of the ties that keep it together and break it apart. This type of wonder emerges when "a certainty . . . has only just been established and has not yet lost the expectation of seeing its opposite (re)appear" (Verhoeven 1972: 27). It is an activity, not a condition; it resonates with the act of wondering, keeping options open so that there is room for the unexpected, or even undesirable, to seep through and enrich our sense-making. In this form, wonder is related to what Bernard of Clairvaux called *admirable mixturae*, those events or phenomena in which ontological and moral boundaries are crossed, confused, or erased (Bynum 1997: 21). The devices I have studied perform such mixtures, all the while giving the impression of being unremarkable technocratic artifacts. To me, they incite all sorts of curiosities; they are microcosms that open doors to make sense of how concept, matter, and affect take form. When guided by this sense of wonder, one opens space to ponder without rushing to final answers. The urgency of the challenges we face demands nothing less. I hope that this sense of wonder has been contagious. A lot hangs on the devices that help run our lives; we need to look at them with generous, critical, and open eyes. Perhaps, as I mentioned at the beginning of this book, at a time of crisis, what we need is to turn to wonder.

NOTES

PREFACE

1. Recent anthropological works describe wonder as "an index of ontological crisis and transformation" (Scott 2013: 860). In this usage, wonder remains tied to the fantastic, to that which is difficult to imagine from the lifeworld of the anthropologist. My project is to take wonder out of that context and examine its epistemic and ontological possibilities without necessitating the existence of the fantastic as its precondition.

2. There is a strong parallel here with the effects that the recognition of global warming has had on Euro-America, a peculiar end of anthropocenic times.

3. The full title of the book is *The Fardle of Façions, conteining the anunciente manners, customes and laws of the peoples enhabiting the two partes of the earth, called Affrike and Asie.*

4. In anthropology, collecting customs and manners led to the fundamental methodology by which the larger comparative project on which the discipline was built would later unfold. James George Frazer (1935), one of anthropology's greatest collectors, put together an outstanding number of customs, practices, rituals, and institutions to illustrate what at the time were thought of as evolutionary patterns. Needless to say, we continue to struggle with the legacy of evolutionary ideas and colonial assumptions undergirding this anthropological legacy, but what remains interesting to me is the analytic presumptions that went into Frazer's collecting drive. Frazer's ever-expanding book, *The Golden Bough: A Study in Magic and Religion*, first published in 1890, consists of sixty-nine chapters densely packed with customs and manners—his chapter titles include "The King of the Wood," "The Magical Control of the Weather," "The Worship of Trees," "Tabooed Persons," "The Myth of Adonis," "Homeopathic Magic of a Flesh Diet," "The Transference of Evil," and so on. Together, the chapters create a dense field where the reader can move laterally between customs, finding her own connections

and disjunctures. *The Golden Bough* invites readers to renounce any intention of mastering content and instead opens the door to linger in wonder. Rather than setting up vertical relationships between an element and the broader complex cultural milieu to which it belonged, as Franz Boas would later do, Frazer was invested in the particularity of each item and its connection to a larger argument about history. He designed a book with analogical relations in mind.

INTRODUCTION

1. This does not imply that they are misguided by the desire to purify the world, as Latour has described it. Theirs is an effort to act ethically amid the difficulties of finding clear courses of action that reflect their commitments. They want to enact distinctions in order to make ethical options possible.

2. These bifurcations are not perfectly symmetric. Different branches have different weights and histories. I want to thank Jörg Niewöhner and the Humboldt STS lab for making this observation.

3. There is a long tradition in science and technology studies of thinking with devices, particularly material ones. Most recently, Law and Ruppert have expanded the reach of the concept to think of devices as lively, unpredictable, and tactical arrangements. There is a lot shared between my conception of the device and Law and Ruppert's, although I am particularly interested in noting the device's role as makers of separations, rather than emphasizing the fact that they are "social"—that is, that they are relational (Law and Ruppert 2013).

4. Michel Foucault's notion of *dispositif* as a tangle of lines of continuity and disruption of power, knowledge, and subject formation has also been theorized as a device (Callon 1998). Foucault's dispositif is often interpreted as operating through a form of synecdoche to the extent that it is capable of standing for an already accomplished epochal configuration. The notion of device that I examine in this book is not as concerned, at least for the time being, with its capacity for epochal diagnosis. Maybe it is more like a cell phone, which as we have seen, despite its apparently limited original significance, has radically reshaped social, financial, and material relations. Giorgio Agamben takes on this notion of the dispositif/device, which he translates as an apparatus, and expands it by noting, "I shall call an apparatus literally anything that has in some way the capacity to capture, orient, determine, intercept, model, control, or secure the gestures, behaviors, opinions, or discourses of living beings. Not only, therefore, prisons, madhouses, the panopticon, schools, confession, factories, disciplines, judicial measures, and so forth (whose connection with power is in a certain sense evident), but also the pen, writing, literature, philosophy, agriculture, cigarettes, navigation, computers, cellular telephones and—why not—language itself, which is perhaps the most ancient of apparatuses—one in which thousands

and thousands of years ago a primate inadvertently let himself be captured, probably without realizing the consequences that he was about to face" (Agamben 2009: 14).

5. The expression *an iota of difference* comes from the early history of Christianity. During its early years, when the separation or embeddedness of church within state was being elucidated, bishops and politicians differed in regard to the nature of Christ and hence the legitimacy of his and the apostles' teachings as Godly word. The controversy was based on the letter *iota* in the Greek alphabet (ι). When describing the relationship between Christ and God, some favored the use of the word *homo-i-ousios*, meaning that the son, Christ, was similar to but not the same as his father, God. The opposing group held that the word *homo-ousios*, meaning one in being or one and the same, described the relation between Father and Son, making Christ one and the same with God itself. The political implications were great, as this determined the relationship between the emperor and Christian representatives on earth, whether they were embedded into each other or not. I want to thank Andrew Mathews for suggesting thinking about the difference that a small difference can make in these terms.

6. I am borrowing Donna Haraway's notion of material-semiotic to keep alive the layered ontology of nonhuman beings in a way that might become invisible when using terms from the "new materialism" turn in the human sciences. The notion of the material-semiotic attunes us to material presence without erasing the semiotic preconditions, inequalities, capacities, and consequences that make their being possible.

7. There is a deep affinity between this approach and STS lineages that have called our attention to the material liveliness of scientific accomplishments and controversies, namely what is glossed as actor–network theory. But there is also a deep affinity between my approach and a variety of older anthropological approaches that found the world was materially constituted through objects that did not fit the Euroamerican categories that analysts assumed when studying the organization of collective life elsewhere.

8. For a recent example see Tsing (2013).

9. This is a task for which ethnography is suitably equipped. Winthereik and Verran speak about its possibilities when they analyze ethnographic stories in terms of part/whole and one/many configurations (Winthereik and Verran 2012).

10. This has also been theorized by Bergson and Deleuze as the virtual.

11. The World Economic Forum brings together private sector, banking, and state representatives to discuss the outlook of the world economy and analyze trends and risks.

12. The report is produced from a survey of 750 "decision-makers and experts" from the forum's "multi-stakeholder communities," and is taken as a gauge of the crises that preoccupy the global political and economic establishment.

1 FORMULA

1. I consider the technical as a set of ideas, instruments, and materials organized around means–ends logics and following a specific set of epistemic orientations with rules and standards that are claimed as particular to a certain field.

2. These general principles of price regulation are not peculiar to Costa Rica—thus, my focus on economic regulation as a set of practices that are at once globalized and contextual. In this chapter, I follow an ethnographic approach that considers the contingencies of locality and the principles of technical abstraction together. For a discussion of the tension between particularity and technical generality in regard to calculation, see Miller (2008) and Appadurai (2013).

3. Since then, ARESEP has moved its headquarters further west in San José, the area where most commercial developments were happening at the time. The building it now occupies is shielded by glass, its lobby is all white and open, and next to it are the offices of a private bank.

4. AyA, a public utility owned by the Costa Rican state, provides water access to close to 60 percent of the country's population. The rest of the country is supplied by ASADAs (community aqueducts) under the legal supervision of AyA, or by municipalities.

5. After a reorganization of ARESEP, WED was transformed into the Intendencia de Aguas. Since the fieldwork for this chapter was conducted while WED was still in existence, I refer to the team with that acronym.

6. On framing as a device for ontological multiplication rather than reduction, see Hetherington (2014).

7. During this period, the share of public investment in the economy rose from 21 percent in the 1960s to 40 percent in the 1980s (Martínez Franzoni and Diego Sánchez-Ancochea 2013: 154). This regime provided benefits to citizens across economic classes, which led to the rapid accumulation of capital and transformed a mainly agricultural economy into one dominated by computer microchip exports and tourism (Vargas Solis 2011).

8. Electricity and water services had been regulated since the 1920s, but until the creation of ARESEP, the regulatory function was performed from within the utilities.

9. In addition to overseeing public utilities, ARESEP regulates the prices of public transportation services (bus and taxi) and oil commercialization, both provided by private companies.

10. An important example that brings the point home is the case of Walmart (Petrovich and Hamilton 2006). As a large buyer of a variety of commodities, Walmart often "imposes" on or, dare I say, regulates the accounting and management practices of its suppliers, very much in the way that regulators oversee

utilities. If one classified ARESEP and Walmart on the basis of their public or private nature, they would seem radically different entities. Yet if one analyzes their price-setting methods and routines, they begin to look similar. This shift in perspective suggests the need to bracket our expectations of what public or private entities do and to ethnographically trace the specific practices that make up contemporary capitalist formations. I thank Matthew Hull for pointing out the parallel between regulatory and large-retail cost-and-profit calculation practices.

11. Michel Callon proposes that economics is not a form of knowledge that depicts an already existing state of affairs but a set of instruments and practices that contribute to the construction of economic settings, actors, and institutions (MacKenzie et al. 2007). What economic knowledge claims to merely describe, it in fact helps bring into existence, formatting and shaping its particularities (Callon et al. 2002; Mitchell 2005).

12. I want to thank Martha Poon for this observation.

13. The *colon* is Costa Rica's currency; *colones* is its plural.

14. The Salamanca School was the intellectual and religious home of Francisco de Vitoria and Bartolome de las Casas, religious thinkers who today are recognized as the precursors of doctrines of universal rights that preceded human rights declarations.

15. Martin Hayek was an Austrian economist recognized as one of the fathers of what is popularly understood as "neoliberal" economics. Hayek argued that freedom resided in unleashing free market dynamics to coordinate social needs and the distribution of wealth. He emphasized the importance of prices as social accomplishments and instruments for coordination.

16. Heredia is the second largest city in Costa Rica.

17. The law that did so, however, framed the new market under the principles of solidarity and universality in service provision. The specific implications of these principles are still being elucidated.

18. It is striking that those two countries were among those that implemented the "neoliberal" economic program more aggressively in the 1990s and 2000s. Chile is the only country in the world to have a private market for water rights, and Argentina undertook a massive privatization policy that turned many utilities into transnational property. Notably, in 2005, Aguas Argentinas, a Suez subsidiary with a concession to manage water infrastructures in Buenos Aires, lost its concession. The company was demanding a price increase and the government requested infrastructural improvements before any augmentation. The company did not accept the conditions, and the government proceeded to end the concession. In 2017, however, under international private arbitration, Argentina was forced to pay a $384 million fine for that decision.

2 INDEX

1. For another approach to the life of material objects as humanitarian traces, see Peter Redfield (2016).

2. For an analysis of the adequacy and acceptance of paying for water on the basis of localized notions of value, see Page (2005).

3. The Bolivian "water war" is an iconic historical moment of water activism. It is invoked as an example of greed, injustice, and, more importantly, immoral behavior by corporations.

4. Since then ARESEP has been completely reorganized. Water services are now under the jurisdiction of the Water Superintendency, a semi-autonomous division within ARESEP whose director has more discretion over the technical decisions they make than the WED director ever had.

5. The inflation rate has become so central to economic activity that central banks use it as a way to shape the future through the periodic definition of inflation goals that they publicly announce. For a thorough analysis of the cultural practices behind this process, see Holmes (2009).

6. Among those changes, we find that in 2005, for instance, the Encuesta included 5,420 households located in the country's metropolitan area—a region where 45 percent of the country's population and 59.3 percent of the country's consumption expenses are concentrated (Instituto Nacional de Estadística y Censos 2006). During its last iteration in 2015, teams of interviewers expanded their scope and went to urban centers outside the metropolitan area following recommendations by the World Bank, the International Monetary Fund, and the Economic Commission for Latin America (CEPAL) to make their national statistics more robust. As a result, this survey reached almost seven thousand households, covering 73 percent of the population and 82 percent of the country's consumption universe.

7. Metamorphoses of the human in relation to the household and the CPI.

8. For another account of how counting makes and unmakes subjects, see Nelson (2015).

9. Costa Rica's first census was produced in 1864 by the then recently created National Statistics Office.

10. These are all offered using the diminutive form of the noun. A polite way of welcoming your guest into your space, the diminutive is a speech form that characterizes much of Costa Rican and Brazilian talk.

3 LIST

1. In the humanities and social sciences there has been a "new materialist" turn that has discovered how the material properties of the worlds people in-

habit shape not only our sense-making but our precognitive worldly possibilities. This turn comes after previous engagements with materiality in anthropology, science and technology studies, and other fields including Marxist, feminist, and material culture theories. The new materialist turn claims to access matter on its own terms while relying heavily on scientific accounts of being that STS and anthropology of science scholars have shown are never "objective" accounts of the material substrate or devoid of human passion and meaning-making.

2. The judicialization of politics consists of the systematic use of the judiciary as a site where fundamental questions about the organization of society, economy, and its relation to "nature" are elucidated. For an expanded analysis of this phenomenon, see Couso (2010), Randeria (2007), and Sikkink (2005).

3. In Costa Rica this was also an outcome of the creation in 1989 of a special constitutional court within the country's supreme court.

4. The name for this category of goods comes from the figure of the *pater*, the male head of the Roman household.

5. José Merino del Río passed away at an early age after undergoing cancer-related surgery in Cuba in 2012.

6. The popular discontent that the Movimiento Libertario had previously captured found a new channel in the 2018 election. Using homophobic, xenophobic, tough on crime, and economically liberal ideas to take advantage of a unique conjuncture, a neopentecostal evangelical party secured second place in the country's presidential elections. While the decision of the Interamerican Human Rights Court to recognize full equality for gay citizens, including the right to marry, ignited the conservative streak of many Costa Ricans, what the election revealed was the wide expansion of evangelical religious infrastructures throughout the country. This electoral development is not an isolated occurrence. The United States, Guatemala, Brazil, Mexico, and Peru are also experiencing an incursion of religion into liberal politics.

7. Libertarian and liberal groups across the continent are linked through institutions such as the Cato Institute in the United States, the Friedrich Naumann Foundation in Germany, and the Latin American Liberal Network, headquartered in Mexico City. They organize training, research, and conferences to strengthen "liberal" thinking and reach new and younger potential members.

8. The Libertarian anti-immigration discourse is targeted toward people from Nicaragua, who are the largest immigrant group in the country. By 2011, Nicaraguan immigrants accounted for about 6 percent of the country's population (Sandoval-Garcia 2015).

9. Since 1956, political parties in Costa Rica that secure more than 4 percent of the national or provincial vote if competing for congressional seats, and not for presidential elections, are entitled to financial support from the state for their campaigns (Sobrado González 2009).

10. One can easily see the parallels between the Libertarian literalism and a Protestant form of biblical literalism. See Crapanzano (2000) and Harding (2000).

11. Borges's encyclopedia is used by Michel Foucault in his preface to *The Order of Things*.

12. Anthropological history is full of prominent lists. Early on, Malinowski, for instance, collected lists of spells, investigating their wording and itemizing structure to explain "the prosaic pedantry of magic" and its dependence on repetition as a way to expand the potency of the utterance (Tambiah 1968: 192). In his argument challenging the dichotomous classification of literate and illiterate societies, Jack Goody (1977) examines lists as a way to enter the systemic logic behind classificatory practices among illiterate societies. He considers their compilation a fundamental way of creating knowledge at the very foundational levels of politics, history, and semiosis. And more recently, Levi-Strauss's (1955) curiosity about the fundamental principles of meaning-making was investigated through the organization of lists of myths and their relations.

13. I want to thank Paul Kockelman for offering guidance in exploring the rich linguistic anthropology literature on lists.

4 PACT

1. *Técnicos* is the word used to refer to technocratic cradles. It signals that, although embedded in public institutions, these men and women are not openly involved in electoral politics but instead draw their influence from the technical knowledge they hold and produce.

2. Bulk water is water directly extracted from reservoirs and canals by users themselves. WMC explains charges for water use as charges for water itself, not for its transportation or treatment.

3. Up to the present, the question of who really pays for these charges remains a controversial one, and while there have been improvements in the effective implementation of the charge system, many contend that medium and small agribusinesses always find ways to avoid paying the full amount they should. Large multinational corporations like Dole have subscribed to the program and are one of the main sources of income for the agency that collects bulk water charges.

4. Northeastern Brazil holds the second worst human development indices in the country, surpassed only by Amazonia.

5. Ernesto's son's election as governor of Ceará in 2015 is another example of how kin relations continue to determine political status and access to state institutions.

6. For an analysis of concrete as a material for future making, see also Nicholas D'Avella (2019).

7. On the life of refusal as a political strategy, see Audra Simpson (2014).

8. For a different take on the role of Post-its, see Wilf (2016).

9. During the heyday of structuralist thought, anthropologists examined aggregation to elucidate the genesis of social collectives that shared "culture or ethnicity." The concept of aggregation remained too tied up in questions of social structure and was soon abandoned as both empirical fact and social explanation (Guyer 1999).

10. I want to thank Katie Ulrich for suggesting this interpretation of the tension between histories.

CONCLUSION

1. Cenotes are natural wells of different dimensions that resemble underground water caves and can be accessed from the surface.

2. For instance, Werth (2016, 2018) traces how, in penal institutions in the United States, assessment instruments and classifications shape the everyday practice of correction, and combine institutional legacies, philosophies of risk, and the affective liveliness of the individuals responsible for penal subjects.

3. Within the realm of law, Silbey (2011) has theorized this kind of person using the figure of the "sociological citizen," a subject that endures the responsibility of implementing, policing, or adapting norms and rules designed to fulfill an image of what society should be. These subjects include judges, inspectors, certifiers, etc.

4. For a broader take on becoming, inspired by Deleuzian thought, see Biehl and Locke (2017).

REFERENCES

Abram, Simone, and Gisa Weszkalnys. 2013. "Elusive Promises: Planning in the Contemporary World. An Introduction." In *Elusive Promises: Planning in the Contemporary World*, edited by Simone Abram and Gisa Weszkalnys, 1–34. New York: Berghahn.

Agamben, Giorgio. 2009. *What Is an Apparatus? And Other Essays*. Stanford, CA: Stanford University Press.

Agamben, Giorgio. 2011. *The Kingdom and the Glory: For a Theological Genealogy of Economy and Government*. Stanford, CA: Stanford University Press.

Alexander, Catherine. 2017. "The Meeting as Subjunctive Form: Public/Private IT Projects in British and Turkish State Bureaucracies." *Journal of the Royal Anthropological Institute* 23, no. S1: 80–94.

Anand, Nikhil. 2017. *Hydraulic City: Water and the Infrastructures of Citizenship in Mumbai*. Durham, NC: Duke University Press.

Anderson, Benedict. 1991. *Imagined Communities: Reflections on the Origins and Spread of Nationalism*. London: Verso.

Ansell, Aaron. 2014. *Zero Hunger: Political Culture and Antipoverty Policy in Northeast Brazil*. Chapel Hill: University of North Carolina Press.

Appadurai, Arjun. 1986. "Introduction: Commodities and the Politics of Value." In *The Social Life of Things: Commodities in Cultural Perspective*, edited by Arjun Appadurai, 3–63. Cambridge: Cambridge University Press.

Appadurai, Arjun. 2013. *The Future as Cultural Fact: Essays on the Global Condition*. London: Verso.

Arendt, Hannah. 1959. *The Human Condition: A Study of the Central Dilemmas Facing Modern Man*. Garden City, NY: Anchor.

Asociación de Entes Reguladores de Agua Potable y Saneamiento de las Américas. 2007. "Esquemas Tarifarios, Regulación y Estructura de las Tarifas. Unidad 2." In *Programa de Teleformación en Regulación de Servicios Publicos 2007*. Asunción: ADERASA.

Assembléia Legislativa do Estado do Ceará, and Conselho de Altos Estudos e Assuntos Estratégicos. 2009. *Plano Estratégico dos Recursos Hídricos do Ceará*. Fortaleza: INESP.

Bakker, Karen. 2007. "The 'Commons' versus the 'Commodity': Alter-Globalization, Anti-Privatization and the Human Right to Water in the Global South." *Antipode* 39, no. 3: 430–55. doi: 10.1111/j.1467-8330.2007.00534.x.

Balbus, Isaac. 1977. "Commodity Form and Legal Form: An Essay on the 'Relative Autonomy' of the Law." *Law and Society Review* 11:571–88.

Balibar, Étienne. 2004. "Is a Philosophy of Human Civic Rights Possible? New Reflections on Equaliberty." *South Atlantic Quarterly* 103, no. 2/3: 311–22.

Ballestero, Andrea. 2004. "Institutional Adaptation and Water Reform in Ceará: Revisiting Structures for Social Participation at the Local Level." Master's thesis, School of Natural Resources and Environment, University of Michigan, Ann Arbor.

Ballestero, Andrea. 2006. "Construcción del Espacio Político a Través de las Prácticas Locales: Bajo Jaguaribe y la Política de Recursos Hídricos." In *Economía Política da Urbanização do Baixo Jaguaribe*, edited by Denise Elias, Renato Pequeno, and Edilson Pereira Jr. Fortaleza: UECE.

Ballestero, Andrea. 2012. "The Productivity of Non-Religious Faith: Openness, Pessimism and Water in Latin America." In *Nature, Science and Religion*, edited by Catherine Tucker, 169–90. Santa Fe, NM: School for Advanced Research.

Ballestero, Andrea. 2019. "Underground as Infrastructure? Figure/Ground Reversals and Dissolution in Sardinal." In *Infrastructure, Environment and Life in the Anthropocene*, edited by Kregg Hetherington, 18–44. Durham, NC: Duke University Press.

Bakker, Karen. 2003. *An Uncooperative Commodity: Privatizing Water in England and Wales*. Oxford: Oxford University Press.

Barad, Karen. 2003. "Posthumanist Performativity: Toward an Understanding of How Matter Comes to Matter." *Signs* 28, no. 3: 801–31.

Barbalho, Alexandre. 2007. "Os Modernos e Os Tradicionais: Cultural Politica no Ceará Contemporâneo." *Estudos de Sociologia* 12, no. 22: 27–42.

Bear, Laura. 2014. "Doubt, Conflict, Mediation: The Anthropology of Modern Time." *Journal of the Royal Anthropological Institute* 20: 3–30. doi: 10.1111/1467-9655.12091.

Benton, Adia. 2015. *HIV Exceptionalism: Development through Disease in Sierra Leone*. Minneapolis: University of Minnesota Press.

Bergson, Henri. 2002. "Concerning the Nature of Time." In *Henri Bergson: Key Writings*, edited by Keith Ansell Pearson and John Mullarkey, 205–19. New York: Continuum.

Biehl, João, and Peter Locke. 2017. *Unfinished: The Anthropology of Becoming*. Durham, NC: Duke University Press.

Bluemel, Eric B. 2004. "The Implications of Formulating a Human Right to Water." *Ecology and Law Quarterly* 31: 957–1006.

Boelens, Rutgerd, and Margreet Zwarteveen. 2005. "Prices and Politics in Andean Water Reforms." *Development and Change* 36, no. 4: 735–58. doi: 10.1111/j.0012–155X.2005.00432.x.

Braidotti, Rosi. 2008. "In Spite of the Times: The Postsecular Turn in Feminism." *Theory, Culture and Society* 25, no. 6: 1–24.

Bynum, Caroline Walker. 1997. "Presidential Address: Wonder." *American Historical Review* 102, no. 1: 1–26.

Callon, Michel. 1998. *The Laws of the Market*. Oxford: Blackwell.

Callon, Michel. 2007. "What Does It Mean to Say That Economics Is Performative?" In *Do Economists Make Markets? On the Performativity of Economics*, edited by Donald MacKenzie, Fabian Muniesa, and Lucia Siu, 311–57. Princeton, NJ: Princeton University Press.

Callon, Michel, Cécile Méadel, and Vololona Rabeharisoa. 2002. "The Economy of Qualities." *Economy and Society* 31, no. 2: 194–217.

Candea, Matei, Joanna Cook, Catherine Trundle, and Thomas Yarrow. 2015. *Detachment: Essays on the Limits of Relational Thinking*. Manchester, UK: Manchester University Press.

Caprara, Andrea, José Wellington de Oliveira Lima, Alice Correia Pequeno Marinho, Paola Gondim Calvasina, Lucyla Paes Landim, and Johannes Sommerfeld. 2009. "Irregular Water Supply, Household Usage and Dengue: A Bio-Social Study in the Brazilian Northeast." *Cadernos de Saúde Pública* 25, no. 1: S125–36.

Carvalho, José Murilo de. 1997. "Mandonismo, Coronelismo, Clientelismo: Uma Discussão Conceitual." *DADOS: Revista de Ciencias Sociais* 40, no. 2.

CEPAL. 2013. *Anuario Estadístico de América Latina y el Caribe*. Santiago, Chile: CEPAL.

Cetina, Karin Knorr. 2006. "The Market." *Theory, Culture and Society* 23, no. 2–3: 551–56. doi: 10.1177/0263276406062702.

Chakrabarty, Dipesh. 2000. *Provincializing Europe: Postcolonial Thought and Historical Difference*. Princeton, NJ: Princeton University Press.

Chen, Cecilia, Janine MacLeod, and Astrida Neimanis, eds. 2013. *Thinking with Water*. Montreal: McGill-Queen's University Press.

Chilcote, Ronald H. 1990. *Power and the Ruling Class in Northeast Brazil*. Cambridge: Cambridge University Press.

Coddington, Mark A. 2015. "Telling Secondhand Stories: News Aggregation and the Production of Journalistic Knowledge." PhD diss., School of Journalism, University of Texas at Austin.

Cohen, Daniel Aldana. 2016. "The Rationed City: The Politics of Water, Housing, and Land Use in Drought-Parched São Paulo." *Public Culture*, no. 28: 261–89. doi: 10.1215/08992363-3427451.

Collier, Jane F., Bill Maurer, and Liliana Suarez-Navaz. 1997. "Sanctioned Identities: Legal Constructions of Modern Personhood." *Identities* 2, no. 1–2: 1–27.
Costales, Raul. 2002. "Progress in Costa Rica." *Liberty Magazine*, 40–41.
Couso, Javier. 2010. "The Transformation of Constitutional Discourse and the Judicialization of Politics in Latin America." In *Cultures of Legality: Judicialization and Political Activism in Latin America*, edited by Javier Couso, Alexandra Huneeus, and Rachel Sieder, 141–60. Cambridge: Cambridge University Press.
Crapanzano, Vincent. 2000. *Serving the Word: Literalism in America from the Pulpit to the Bench*. New York: New Press.
Daston, Lorraine, and Katharine Park. 1998. *Wonders and the Order of Nature, 1150–1750*. New York: Zone.

D'Avella, Nicholas. 2019. *Concrete Dreams: Practice, Value, and Built Environments in Post-Crisis Buenos Aires*. Durham, NC: Duke University Press.
De Goede, Marieke. 2005. *Virtue, Fortune, and Faith: A Genealogy of Finance*. Minneapolis: University of Minnesota Press.
de la Bellacasa, Maria Puig. 2011. "Matters of Care in Technoscience: Assembling Neglected Things." *Social Studies of Science* 41, no. 1: 85–106.
de la Cadena, Marisol. 2015. *Earth Beings: Ecologies of Practice across Andean Worlds*. Durham, NC: Duke University Press.
de Oliveira, Marcilio Caetano. 2008. *Modelos de Alocação e Realocação de Água: Um Estudo de Caso do Programa "Águas do Vale" nos Rios Jaguaribe e Banabuiú*. Fortaleza: Department of Environmental and Hydraulic Engineering, Universidade Federal do Ceará, Fortaleza, CE, Brazil.
Descola, Philippe. 2013. *The Ecology of Others*. Translated by Geneviève Godbout and Benjamin P. Luley. Chicago: Prickly Paradigm.
do Amaral Filho, Jair. 2003. *Reformas Estruturais e Economia Politica dos Recursos Hidricos no Ceara*. Fortaleza: SEPLAN-IPECE.
Edwards, Paul N. 2010. *A Vast Machine: Computer Models, Climate Data, and the Politics of Global Warming*. Cambridge, MA: MIT Press.
Elegido, Juan Manuel. 2009. "The Just Price: Three Insights from the Salamanca School." *Journal of Business Ethics* 90, no. 1: 29–46.
Empson, William. 1977. *The Structure of Complex Words*. New York: Vintage.
Escobar, Miguel. 1972. "Ad-Mirar." In *Dicionário Paulo Freire*, edited by Danilo R. Streck, Euclides Redin, and Jaime José Zitkoski. Belo Horizonte, Brazil: Autêntica Editora.
Evans-Pritchard, E. E. 1940. *The Nuer: A Description of the Modes of Livelihood and Political Institutions of a Nilotic People*. Oxford: Oxford University Press.
Evelyn, Sir George Shuckburgh. 1798. "An Account of Some Endeavours to Ascertain a Standard of Weight and Measure." *Philosophical Transaction of the Royal Society of London*, no. 88: 133–82.

Fassin, Didier. 2012. *Humanitarian Reason: A Moral History of the Present*. Berkeley: University of California Press.

Faubion, James D. 2010. "From the Ethical to the Themitical (and Back): Groundwork for an Anthropology of Ethics." In *Ordinary Ethics: Anthropology, Language, and Action*, edited by Michael Lambek, 84–101. New York: Fordham University Press.

Finkelstein, Andrea. 2000. *Harmony and the Balance: An Intellectual History of Seventeenth-Century English Economic Thought*. Ann Arbor: University of Michigan Press.

Fleming, Sam. 2017. "Fed Has No Reliable Theory of Inflation, Says Tarullo." *Financial Times*, October 4, 2017.

Fortun, Kim. 2012. "Ethnography in Late Industrialism." *Cultural Anthropology* 27, no. 3: 446–64. doi: 10.1111/j.1548-1360.2012.01153.x.

Foucault, Michel. 1973. *The Order of Things*. New York: Vintage.

Foucault, Michel. 2009. *Security, Territory, Population: Lectures at the Collège de France 1977–1978*. New York: Picador.

Fraser, Nancy. 2014. "Can Society Be Commodities All the Way Down? Post-Polanyian Reflections on Capitalist Crisis." *Economy and Society* 43, no. 4: 541–58.

Frazer, Sir James George. 1935. *The Golden Bough: A Study in Magic and Religion*. Vol. I. New York: Macmillan.

Furtado, Celso. 1998. *Seca e Poder: Entrevista com Celso Furtado/Entrevistadores Maria da Conceição Tavares, Manuel Correia de Andrade, Raimundo Pereira*. São Paulo: Editora Fundação Perseu Abramo.

Gad, Christopher, and Brit Ross Winthereik. n.d. "Practices of Proximation: Critique and the Experimental Organization." http://www.academia.edu/27669085/Practices_of_Proximation_Critique_and_the_Experimental_Organization.

Garcia, Angela. 2014. "The Promise: On the Morality of the Marginal and the Illicit." *Ethos* 42, no. 1: 51–64.

Gibson-Graham, J. K. 2006. *The End of Capitalism (As We Knew It): A Feminist Critique of Political Economy*. Minneapolis: Minnesota University Press.

Goody, Jack. 1977. *The Domestication of the Savage Mind*. Cambridge: Cambridge University Press.

Greenfield, Gerald M. 1992. "The Great Drought and Elite Discourse in Imperial Brazil." *Hispanic American Historic Review* 72, no. 3: 375–400.

Greenhouse, Carol. 1996. *A Moment's Notice*. Ithaca, NY: Cornell University Press.

Greenhouse, Carol. 2014. "Time's Up, Timed Out: Reflections on Social Time and Legal Pluralism." *Journal of Legal Pluralism and Unofficial Law* 46, no. 1: 141–53. doi: 10.1080/07329113.2014.882102.

Gregory, Chris A. 1982. *Gifts and Commodities.* London: Academic Press.
Grosz, Elizabeth. 2002. "Feminist Futures?" *Tulsa Studies in Women's Literature* 21, no. 1: 13–20.
Gupta, Akhil, and James Ferguson. 1997. "Discipline and Practice: 'The Field' as Site, Method and Location in Anthropology." In *Anthropological Locations: Boundaries and Grounds of a Field Science*, edited by Akhil Gupta and James Ferguson, 1–46. Berkeley: University of California Press.
Guyer, Jane I. 1999. "Anthropology: The Study of Social and Cultural Originality." *African Sociological Review/Revue Africaine de Sociologie* 3, no. 2: 30–53.
Guyer, Jane I. 2004. *Marginal Gains: Monetary Transactions in Atlantic Africa.* Chicago: University of Chicago Press.

Guyer, Jane I. 2007. "Prophecy and the Near Future: Thoughts on Macroeconomic, Evangelical, and Punctuated Time." *American Ethnologist* 34, no. 3: 409–21.
Guyer, Jane I. 2009. "Composites, Fictions, and Risk: Toward an Ethnography of Price." In *Market and Society: The Great Transformation Today*, edited by Chris Hann and Keith Hart, 203–20. Cambridge: Cambridge University Press.
Guyer, Jane I. 2013. "Indexing People to Money: Beyond Consumption?" Paper presented at the Human Economy Conference, Pretoria, South Africa, August 2013.
Hamouda, O. F., and B. B. Price. 1997. "The Justice of the Just Price." *European Journal of the History of Economic Thought* 4, no. 2: 191–216. doi: 10.1080/10427719700000036.
Haraway, Donna. 1992. "The Promises of Monsters: A Regenerative Politics for Inappropriate/d Others." In *Cultural Studies*, edited by Lawrence Grossberg, Cory Nelson, and Paula A. Treichler, 295–337. New York: Routledge.
Hardin, Garret. 1968. "The Tragedy of the Commons." *Science* 162: 1243–48.
Harding, Susan. 2000. *The Book of Jerry Falwell: Fundamentalist Language and Politics.* Princeton, NJ: Princeton University Press.
Hardt, Michael, and Antonio Negri. 2005. *Multitude: War and Democracy in the Age of Empire.* New York: Penguin.
Hart, Keith. 2007. "Money: Towards a Pragmatic Economic Anthropology." *Memory Bank*, July 15. http://www.thememorybank.co.uk/2007/07/15/127/.
Hayden, Cori. 2003. *When Nature Goes Public: The Making and Unmaking of Bioprospecting in Mexico.* Princeton, NJ: Princeton University Press.
Helgason, Agnar, and Gísli Pálsson. 1997. "Contested Commodities: The Moral Landscape of Modernist Regimes." *Journal of the Royal Anthropological Institute* 3, no. 3: 451–71.
Helmreich, Stefan. 2015. *Sounding the Limits of Life: Essays in the Anthropology of Biology and Beyond.* Princeton, NJ: Princeton University Press.
Henare, Amiria, Martin Holbraad, and Sari Wastell. 2007. "Introduction: Think-

ing through Things." In *Thinking through Things: Theorising Artefacts Ethnographically*, edited by Amiria Henare, Martin Holbraad, and Sari Wastell, 1–31. London: Routledge.

Hepburn, Ronald W. 1980. "The Inaugural Address: Wonder." *Proceedings of the Aristotelian Society*, Supplementary Volumes 54: 1–23.

Hetherington, Kregg. 2014. "Regular Soybeans: Translation and Framing in the Ontological Politics of a Coup." *Indiana Journal of Global Legal Studies* 21, no. 1: 55–78.

Hobbes, Thomas. [1651] 1991. *Leviathan*. Cambridge: Cambridge University Press.

Hodgen, Margaret. 1964. *Early Anthropology in the Sixteenth and Seventeenth Centuries*. Philadelphia: University of Pennsylvania Press.

Holmes, Douglas R. 2009. "Economy of Words." *Cultural Anthropology* 24, no. 3: 381–419.

Instituto Nacional de Estadística y Censos. 2006. *Principales Características del Êndice de Precios al Consumidor: Base Julio 2006*. San José: INEC.

Instituto Nacional de Estadística y Censos. 2016. "Items in the 2015 Consumer Price Index Consumption Basket." San José: INEC.

Jakobson, Roman. 1960. "Linguistics and Poetics." In *Style in Language*, edited by Thomas A. Sebeok, 350–77. Cambridge, MA: MIT Press.

Jasanoff, Sheila. 2011. "Constitutional Moments in Governing Science and Technology." *Science and Engineering Ethics* 17, no. 4: 621–38.

Johns, Fleur. 2016. "Global Governance through the Pairing of List and Algorithm." *Environment and Planning D: Society and Space* 34, no. 1: 126–49.

Jouravlev, Andrei. 2001. *Regulación de la Industria de Agua Potable*, vol. 2: *Regulación de las Conductas*. Santiago: CEPAL.

Juris, Jeffrey S. 2012. "Reflections on #Occupy Everywhere: Social Media, Public Space, and Emerging Logics of Aggregation." *American Ethnologist* 39, no. 2: 259–79.

Keane, Webb. 2010. "Minds, Surfaces, and Reasons in the Anthropology of Ethics." In *Ordinary Ethics: Anthropology, Language, and Action*, edited by Michael Lambek. New York: Fordham University Press.

Kendall, Maurice George. 1969. "Studies in the History of Probability and Statistics, XXI. The Early History of Index Numbers." *Review of the International Statistical Institute* 37, no. 1: 1–12.

Kockelman, Paul. 2013. "The Anthropology of an Equation: Sieves, Spam Filters, Agentive Algorithms, and Ontologies of Transformation." *HAU: Journal of Ethnographic Theory* 3, no. 3: 33–61.

Kopytoff, Igor. 1986. "The Cultural Biography of Things: Commoditization as a Process." In *The Social Life of Things: Commodities in Cultural Perspective*, edited by Arjun Appadurai, 64–91. Cambridge: Cambridge University Press.

Kottak, C., A. Costa, and R. Prado. 1996. "Popular Participation in Brazil: Northeast Rural Development Program." Discussion Paper No. 333. Washington, DC: World Bank.

Latour, Bruno. 1993. *We Have Never Been Modern*. Cambridge, MA: Harvard University Press.

Latour, Bruno. 2005. *Reassembling the Social: An Introduction to Actor-Network-Theory, Clarendon Lectures in Management Studies*. Oxford: Oxford University Press.

Law, John, and Evelyn Ruppert. 2013. "The Social Life of Methods: Devices." *Journal of Cultural Economy* 6, no. 3: 229–40. doi: 10.1080/17530350.2013.812042.

Lemos, Maria Carmen, and João Lúcio Farias de Oliveira. 2004. "Can Water Reform Survive Politics? Institutional Change and River Basin Management in Ceará, Northeast Brazil." *World Development* 32, no. 12: 2121–37.

Lemos, Maria Carmen, Timothy J. Finan, Roger W. Fox, Donald R. Nelson, and Joanna Tucker. 2002. "The Use of Seasonal Climate Forecasting in Policymaking: Lessons from Northeast Brazil." *Climatic Change* 55, no. 4: 479–507.

Levi-Strauss, Claude. 1955. "The Structural Study of Myth." *Journal of American Folklore* 68, no. 270: 428–44.

Li, Tania Murray. 2007. "Practices of Assemblage and Community Forest Management." *Economy and Society* 36, no. 2: 263–93.

Linton, Jamie. 2010. *What Is Water? The History of a Modern Abstraction*. Vancouver: UBC Press.

MacKenzie, Donald A., Fabian Muniesa, and Lucia Siu. 2007. *Do Economists Make Markets?: On the Performativity of Economics*. Princeton, NJ: Princeton University Press.

Malinowski, Bronislaw. 1920. "51. Kula; The Circulating Exchange of Valuables in the Archipelagoes of Eastern New Guinea." *Man* 20: 97–105.

Malinowski, Bronislaw. 1935. *Coral Gardens and Their Magic*, vol. 1. London: Allen and Unwin.

Marcus, George. 2006. "Multi-Sited Ethnography: Five or Six Things I Know about It Now." Paper read at Refunctioning Ethnography, January 27–28, 2006, University of California, Irvine.

Martínez Franzoni, Juliana, and Diego Sánchez-Ancochea. 2013. "Can Latin American Production Regimes Complement Universalistic Welfare Regimes? Implications from the Costa Rican Case." *Latin American Research Review* 48, no. 2: 148–73.

Marx, Karl. 1976. *Capital*, vol. I. London: Penguin Classics.

Masco, Joseph. 2014. *The Theater of Operations: National Security Affect from the Cold War to the War on Terror*. Durham, NC: Duke University Press.

Mathews, Andrew S. 2008. "State Making, Knowledge, and Ignorance: Translation and Concealment in Mexican Forestry Institutions." *American Anthropologist* 110, no. 4: 484–94.

Mathews, Andrew S., and Jessica Barnes. 2016. "Prognosis: Visions of Environmental Futures." *Journal of the Royal Anthropological Institute* 22, no. S1: 9–26. doi: 10.1111/1467-9655.12391.

Mathur, Nayanika. 2016. *Paper Tiger: Law, Bureaucracy, and the Developmental State in Himalayan India*. Cambridge: Cambridge University Press.

Mattern, Shannon. Forthcoming. "The Spectacle of Data: A Century of Fiches, Fairs, and Fantasies." *Theory, Culture and Society*.

Maurer, Bill. 1997. "Colonial Policy and the Construction of the Commons: An Introduction." *Plantation Society in the Americas* 4, no. 2–3: 113–33.

Maurer, Bill. 2005. *Mutual Life Limited: Islamic Banking, Alternative Currencies, Lateral Reason*. Princeton, NJ: Princeton University Press.

Maurer, Bill. 2012. "Finance 2.0." In *A Handbook of Economic Anthropology*, edited by James G. Carrier, 183–201. Cheltenham, UK: Edward Elgar.

Mauss, Marcel. 1967. *The Gift*. New York: W. W. Norton.

Melé, Domènec. 1999. "Early Business Ethics in Spain: The Salamanca School (1526–1614)." *Journal of Business Ethics* 22, no. 3: 175–89.

Miller, Daniel. 2008. "The Uses of Value." *Geoforum* 39, no. 3: 1122–32.

Mitchell, Timothy. 2005. "The Work of Economics: How a Discipline Makes Its World." *European Journal of Sociology* 45, no. 2: 297–320.

Miyazaki, Hirokazu. 2004. *The Method of Hope*. Stanford, CA: Stanford University Press.

Mokyr, Joel. 2001. *The Gifts of Athena: Historical Origins of the Knowledge Economy*. Princeton, NJ: Princeton University Press.

Mora Portuguez, Jorge, and Vanessa Dubois Cisneros. 2015. *Implementación del derecho humano al agua en América Latina*. Caracas: Corporación Andina de Fomento.

Mosse, David. 2003. *The Rule of Water: Statecraft, Ecology and Collective Action in South India*. Oxford: Oxford University Press.

Muehlebach, Andrea. 2012. *The Moral Neoliberal: Welfare and Citizenship in Italy*. Chicago: University of Chicago Press.

Munn, Nancy D. 1986. *The Fame of Gawa: A Study of Value Transformation in a Massim (Papua New Guinea) Society*. Cambridge: Cambridge University Press.

Munn, Nancy D. 1992. "The Cultural Anthropology of Time: A Critical Essay." *Annual Review of Anthropology* 21: 93–123.

Nader, Laura. 1990. *Harmony Ideology: Justice and Control in a Zapotec Mountain Village*. Stanford, CA: Stanford University Press.

Nafus, Dawn, and Ken Anderson. 2009. "Writing on Walls: The Materiality of Social Memory in Corporate Research." In *Ethnography and the Corporate Encounter: Reflections on Research on and in Corporations*, edited by Melissa Cefkin, 137–57. Oxford: Berghahn.

Neiburg, Federico. 2006. "Inflation: Economists and Economic Cultures in Brazil and Argentina." *Comparative Studies in Society and History* 48: 604–33.

Neiburg, Federico. 2010. "Sick Currencies and Public Numbers." *Anthropological Theory* 10, no. 1: 1–7.

Neimanis, Astrida. 2012. "Hydrofeminism: Or, On Becoming a Body of Water." New York: Cambridge University Press.

Nelson, Diane. 2015. *Who Counts? The Mathematics of Death and Life after Genocide*. Durham, NC: Duke University Press.

Office of the High Commissioner for Human Rights. 2003. *General Comment No. 15: The Right to Water*. E/C.12/2002/11. http://www.refworld.org/pdfid/4538838d11.pdf.

Olson, Mancur. 1971. *The Logic of Collective Action: Public Goods and the Theory of Groups*. Cambridge, MA: Harvard University Press.

Ong, Aihwa, and Stephen J. Collier. 2005. *Global Assemblages: Technology, Politics, and Ethics as Anthropological Problems*. Oxford: Blackwell.

Ostrom, Elinor. 1990. *Governing the Commons: The Evolution of Institutions for Collective Action*. New York: Cambridge University Press.

Otlet, Paul. [1918] 1990. "Transformations in the Bibliographical Apparatus of the Sciences: Repository, Classification, Office of Documentation." In *International Organization and Dissemination of Knowledge: Selected Essays of Paul Otlet*, edited by W. Boyd Rayward. New York: Elsevier.

Otlet, Paul. 1920. *L'Organisation internationale de la bibliographie et de la documentation* [International organization and dissemination of knowledge: Selected essays]. Brussels: Institut International de Bibliographie.

Pacheco-Vega, Raul. 2015. "Agua embotellada en México: De la privatización del suministro a la mercantilización de los recursos hídricos." *Espiral: Estudios sobre Estado y Sociedad* 22, no. 63: 221–63.

Page, Ben. 2005. "Paying for Water and the Geography of Commodities." *Transactions of the Institute of British Geographers* 30, no. 3: 293–306.

Pashukanis, Evgeny. 1980. *Selected Writings on Marxism and Law*. London: Academic Press.

Petrovich, Misha, and Gary G. Hamilton. 2006. "Making Global Markets: Wal-Mart and Its Suppliers." In *Wal-Mart: The Face of Twenty-first Century Capitalism*, edited by Nelson Lichtenstein, 107–41. New York: New Press.

Philips, Andrea. 2012. "List." In *Inventive Methods: The Happening of the Social*, edited by Celia Lury and Nina Wakeford, 96–109. London: Routledge.

Polanyi, Karl. 1957. *The Great Transformation*. Boston: Beacon.

Ponge, Francis, and Beth Archer Brombert. 1972. *The Voice of Things*. New York: McGraw-Hill.

Poovey, Mary. 1998. *A History of the Modern Fact: Problems of Knowledge in the Sciences of Wealth and Society*. Chicago: University of Chicago Press.

Pottage, Alain. 2001. “Persons and Things: An Ethnographic Analogy.” *Economy and Society* 30, no. 1: 112–38.

Potter, Michael. 2004. *Set Theory and Its Philosophy: A Critical Introduction*. Oxford: Oxford University Press.

Povinelli, Elizabeth A. 2011. *Economies of Abandonment: Social Belonging and Endurance in Late Liberalism*. Durham, NC: Duke University Press.

Radin, Margaret Jane. 1996. *Contested Commodities*. Cambridge, MA: Harvard University Press.

Raffles, Hugh. 2002. *In Amazonia: A Natural History*. Princeton, NJ: Princeton University Press.

Randeria, Shalini. 2007. “De-Politicization of Democracy and Judicialization of Politics.” *Theory, Culture and Society* 24, no. 4: 38–44. doi: 10.1177/0263276407080398.

Redfield, Peter. 2016. “Fluid Technologies: The Bush Pump, the LifeStraw® and Microworlds of Humanitarian Design.” *Social Studies of Science* 46, no. 2: 159–83.

Riles, Annelise. 2013. “Market Collaboration: Finance, Culture, and Ethnography after Neoliberalism.” *American Anthropologist* 115, no. 4: 555–69.

Ringel, Felix. 2016. “Beyond Temporality: Notes on the Anthropology of Time from a Shrinking Fieldsite.” *Anthropological Theory* 16, no. 4: 390–412.

Robbins, Joel. 2013. “Beyond the Suffering Subject: Toward an Anthropology of the Good.” *Journal of the Royal Anthropological Institute* 19, no. 3: 447–62. doi: 10.1111/1467-9655.12044.

RobecoSAM. 2015. *Water: The Market of the Future*. Zurich: RobecoSAM AG.

Roberts, Elizabeth. 2017. “What Gets Inside: Violent Entanglements and Toxic Boundaries in Mexico City.” *Cultural Anthropology* 32, no. 4: 592–619.

Roitman, Janet L. 2005. *Fiscal Disobedience: An Anthropology of Economic Regulation in Central Africa*. Princeton, NJ: Princeton University Press.

Roitman, Janet. 2013. *Anti-Crisis*. Durham, NC: Duke University Press.

Rosenberg, Daniel, and Susan Harding. 2005. “Introduction: Histories of the Future.” In *Histories of the Future*, edited by Daniel Rosenberg and Susan Harding, 1–18. Durham, NC: Duke University Press.

Rotman, Brian. 1997. “The Truth about Counting.” https://brianrotman.wordpress.com/articles/1997-the-truth-about-counting/.

Rubenstein, Mary-Jane. 2006. “A Certain Disavowal: The Pathos and Politics of Wonder.” *Princeton Theological Review* 12, no. 2: 11–17.

Ryle, Gilbert. 1945. “Knowing How and Knowing That: The Presidential Address.” *Proceedings of the Aristotelian Society*, n.s. 46: 1–16.

Sandoval-Garcia, Carlos. 2015. “Nicaraguan Immigration to Costa Rica: Tendencies, Policies, and Politics.” LASA *Forum* 44, no. 4: 7–10.

Sawyer, Suzana. 2017. “Modes of Potentiality.” *GeoHumanities* 3, no. 1: 170–77.

Schmidt, Jeremy J. 2017. *Water: Abundance, Scarcity, and Security in the Age of Humanity*. New York: NYU Press.

Scott, Michael W. 2013. "The Anthropology of Ontology (Religious Science?)." *Journal of the Royal Anthropological Institute* 19, no. 4: 859–72.

Sikkink, Kathryn. 2005. "The Transnational Dimension of the Judicialization of Politics in Latin America." In *The Judicialization of Politics in Latin America*, edited by Rachel Sieder, Line Schjolden, and Alan Angell, 263–92. New York: Palgrave Macmillan.

Silbey, Susan S. 2011. "The Sociological Citizen: Pragmatic and Relational Regulation in Law and Organizations." *Regulation and Governance* 5, no. 1: 1–13.

Simpson, Audra. 2014. *Mohawk Interruptus: Political Life across the Borders of Settler States*. Durham, NC: Duke University Press.

Simpson, Larry D. 2003. *Integrated Water Resources Management, Ceara, Brazil*. Washington, DC: World Bank.

Smith, Adam. 1966. *The Theory of Moral Sentiments*. New York: A. M. Kelley.

Sobrado González, Luis Antonio. 2009. "La financiación de los partidos políticos en Costa Rica." *Revista de Derecho Electoral* 8. http://www.tse.go.cr/revista/art/8/Sobrado_Gonzalez.pdf.

Spackman, Christy, and Gary A. Burlingame. 2018. "Sensory Politics: The Tug-of-War between Potability and Palatability in Municipal Water Production." *Social Studies of Science*. doi: 0306312718778358.

Stengers, Isabelle. 2005. "The Cosmopolitical Proposal." In *Making Things Public: Atmospheres of Democracy*, edited by Bruno Latour and Peter Weibel, 994–1003. Cambridge, MA: MIT Press.

Stewart, Kathleen. 2013. "The Achievement of a Life, a List, a Line." In *The Social Life of Achievement*, edited by Nicholas J. Long and Henrietta L. Moore, 31–42. New York: Berghahn.

Strang, Veronica. 2006. "Fluidscapes: Water, Identity and the Senses." *World Views: Global Religions, Culture, and Ecology* 10, no. 2: 147–54.

Strathern, Marilyn. 1988. *The Gender of the Gift*. Berkeley: University of California Press.

Strathern, Marilyn. 1999. *Property, Substance and Effect: Anthropological Essays on Persons and Things*. London: Athlone.

Strathern, Marilyn. 2011. "Binary License." *Common Knowledge* 17, no. 1: 87–103.

Swaab, Peter. 2012. "'Wonder' as a Complex Word." *Romanticism* 18, no. 3: 270–80.

Tambiah, Stanley Jeyaraja. 1968. "The Magical Power of Words." *Man* 3, no. 2: 175–208.

Tendler, Judith. 1997. *Good Government in the Tropics*. Baltimore: Johns Hopkins University Press.

Thomas, Nicholas. 1991. *Entangled Objects: Exchange, Material Culture, and Colonialism in the Pacific*. Cambridge, MA: Harvard University Press.

Thompson, Charis. 2005. *Making Parents: The Ontological Choreography of Reproductive Technologies*. Cambridge, MA: MIT Press.

Thrift, Nigel. 2005. *Knowing Capitalism*. London: SAGE.

Tiersma, Peter M. 2005. "Categorical Lists in the Law." In *Vagueness in Normative Texts*, edited by Vijay K. Bhatia, Jan Engberg, Maurizio Gotti, and Dorothee Heller, 109–30. Bern: Peter Lang.

Tsing, Anna. 2013. "Sorting Out Commodities: How Capitalist Value Is Made through Gifts." *HAU: Journal of Ethnographic Theory* 3, no. 1: 21–43.

Tully, Shawn. 2000. "Water, Water Everywhere." *Fortune*, May 15, 342–54.

United Nations Development Programme. 2008. *Water Policy and Strategy of UNEP*. Geneva: United Nations Development Programme.

Valverde, Mariana. 2009. "Jurisdiction and Scale: Legal 'Technicalities' as Resources for Theory." *Social and Legal Studies* 18, no. 2: 139–57.

Vargas Solis, Luis Paulino. 2011. "Costa Rica: Tercera fase de la estrategia neoliberal. Contradicciones y desafios (2005–2010)." *Rupturas* 1, no. 2: 84–107.

Verhoeven, Cornelis. 1972. *The Philosophy of Wonder*. New York: Macmillan.

Verran, Helen. 2001. *Science and African Logic*. Chicago: University of Chicago Press.

von der Lippe, Peter. 2012. "Recurrent Price Index Problems and Some Early German Papers on Index Numbers: Notes on Laspeyres, Paasche, Drobisch, and Lehr." *Jahrbücher fur Nationalökonomie und Statistik* 233, no. 3: 336–66.

von Schnitzler, Antina. 2016. *Democracy's Infrastructure: Techno-Politics and Protest after Apartheid*. Princeton, NJ: Princeton University Press.

Weiner, Annette B. 1985. "Inalienable Wealth." *American Ethnologist* 12, no. 2: 210–27.

Weiner, Annette B. 1992. *Inalienable Possessions: The Paradox of Keeping while Giving*. Berkeley: University of California Press.

Werth, Robert. 2016. "Individualizing Risk: Moral Judgement, Professional Knowledge and Affect in Parole Evaluations." *British Journal of Criminology* 57, no. 4: 808–27.

Werth, Robert. 2018. "Theorizing the Performative Effects of Penal Risk Technologies: (Re)Producing the Subject Who Must Be Dangerous." *Social and Legal Studies*. doi: 10.1177/0964663918773542.

West, Paige. 2012. *From Modern Production to Imagined Primitive: The Social World of Coffee from Papua New Guinea*. Durham, NC: Duke University Press.

Weston, Kath. 2016. *Animate Planet: Making Visceral Sense of Living in a High-Tech Ecologically Damaged World*. Durham, NC: Duke University Press.

Wiegman, Robyn. 2000. "Feminism's Apocalyptic Futures." *New Literary History* 31, no. 4: 805–25.

Wilf, Eitan. 2016. "The Post-it Note Economy: Understanding Post-Fordist Business Innovation through One of Its Key Semiotic Technologies." *Current Anthropology* 57, no. 6: 732–60.

Wilk, Richard. 2006. "Bottled Water: The Pure Commodity in the Age of Branding." *Journal of Consumer Culture* 6, no. 3: 303–25.

Wilken, Rowan. 2010. "The Card Index as Creativity Machine." *Culture Machine* 11: 7–30.

Willey, Angela. 2016. "A World of Materialisms: Postcolonial Feminist Science Studies and the New Natural." *Science, Technology and Human Values* 41, no. 6: 991–1014. doi: 10.1177/0162243916658707.

Williams, Raymond. 1977. "Structures of Feeling." *Marxism and Literature*, 128–35. Oxford: Oxford University Press.

Winthereik, Brit Ross, and Helen Verran. 2012. "Ethnographic Stories as Generalizations That Intervene." *Science and Technology Studies* 28, no. 1: 37–51.

World Economic Forum. 2016. *The Global Risks Report*. Geneva: World Economic Forum.

Wutich, Amber. 2009. "Water Scarcity and the Sustainability of a Common Pool Resource Institution in the Urban Andes." *Human Ecology* 37, no. 2: 179–92.

INDEX